AF709631

COMMON CORE STATE STANDARDS

Problem-Based Tasks for Mathematics II

WALCH EDUCATION®

1 2 3 4 5 6 7 8 9 10
978-0-8251-7198-7

J. Weston Walch, Publisher
Portland, ME 04103
www.walch.com
Printed in the United States of America
WALCH EDUCATION

Table of Contents

Table of Contents

Functions

Geometry

Table of Contents

Introduction

Welcome to *Common Core State Standards Problem-Based Tasks for Mathematics II.* Use these engaging real-world scenarios to help infuse your mathematics program with a problem-based approach to the knowledge and skills required by the Common Core State Standards for Mathematics.

This collection of tasks addresses all of the Common Core State Standard conceptual categories for high school mathematics:

- Number and Quantity
- Algebra
- Functions
- Modeling
- Geometry
- Statistics and Probability

The tasks support students in developing and using the Mathematical Practices that are a fundamental part of the CCSS. You can implement these tasks flexibly—to walk students through the application of the standard, prior to traditional instruction, or at the end of instruction. The tasks are appropriate for any Mathematics II class, or for other CCSS-based Grade 10 courses.

Each Problem-Based Task is set in a meaningful real-world context to engage student interest and reinforce the relevance of mathematics. Each is tightly aligned to one or more specific standards from the High School CCSS for Mathematics II. Each task combines the specific content of one or more Common Core State Standards with higher-order thinking. Optional coaching questions scaffold the tasks and guide students in solving the problems. Answers and suggested responses to the coaching questions are provided.

Student pages identify the targeted Common Core State Standard(s) and present the problem-solving tasks in familiar and intriguing contexts, and require collaboration, problem solving, reasoning, and communication. You may choose to assign the tasks with little scaffolding (by forgoing use of the coaching questions), or with the series of coaching questions that currently follow each task to lead students through the important steps of the problem. You can also differentiate to meet the needs of individual students by providing coaching questions as appropriate.

We developed these Problem-Based Tasks at the request of math educators and with advice and feedback from mathematics supervisors and teachers. Please let us know how they work in your classroom. We'd love suggestions for improving the tasks, or topics and contexts for creating additional tasks. Visit us at www.walch.com, follow us on Twitter (@WalchEd), or e-mail suggestions to customerservice@walch.com.

Standards Correlations

Common Core State Standards Problem-Based Tasks for Mathematics II is correlated to the Common Core State Standards for high school mathematics. The table that follows lists each task's targeted Common Core State Standard(s), focus, title, and starting page number. The materials are organized in the order in which the standards are listed in the mathematics CCSS, and are grouped by high school conceptual category. Stars (★) indicate modeling standards. To access the full text of the Common Core State Standards for high school mathematics, view the PDF found at http://www.walch.com/CCSS/00001.

CCSS addressed	Task focus	Task title	Page number
Number and Quantity: The Real Number System			
N–RN.1 N–RN.2	Defining, Rewriting, and Evaluating Rational Exponents	Population Growth	1
N–RN.2 N–RN.3	Rational and Irrational Numbers and Their Properties	Estimating Depreciation	4
Number and Quantity: The Complex Number System			
N–CN.1	Defining Complex Numbers, i, and i^2	Representing Impedance	8
N–CN.2	Adding and Subtracting Complex Numbers	Elements in Series in a Circuit	11
N–CN.2	Multiplying Complex Numbers	Elements in Parallel in a Circuit	14
N–CN.7 N–CN.9 (+)	Solving Quadratic Equations with Complex Solutions	All Roads Lead to the Same Destination	17
N–CN.8 (+)	Extending Polynomial Identities to Include Complex Numbers	Making a Complex Expression Simple	22
Algebra: Seeing Structure in Expressions			
A–SSE.1a★	Identifying Terms, Factors, and Coefficients	Deck the Deck	27
A–SSE.1b★	Interpreting Complicated Expressions	Puppy Pen	30
A–CED.2★ A–SSE.3a★	Creating and Graphing Equations Using Standard Form	Parabolic Party Streamers	34

(*continued*)

Standards Correlations

CCSS addressed	Task focus	Task title	Page number
A–CED.2★ A–SSE.3a★	Creating and Graphing Equations Using the x-intercepts	It's Show Time!	37
A–CED.2★ A–SSE.3b★	Creating and Graphing Equations Using Vertex Form	Rocket Heights	41
A–SSE.3c★	Writing Exponential Expressions in Equivalent Forms	Options of Interest	45
Algebra: Arithmetic with Polynomials and Rational Expressions			
A–APR.1	Adding and Subtracting Polynomials	Cabin Perimeter	49
A–APR.1	Multiplying Polynomials	Architectural Area	52
Algebra: Creating Equations			
A–CED.1★	Creating Rational Equations	Golden Rectangles and the Golden Ratio	55
A–CED.1★	Creating Rational Inequalities	Capping Costs	59
A–CED.2★	Graphing Rational Equations	Lacrosse Stats	63
A–CED.4★	Rearranging Formulas	Selecting a Cylinder Size	68
Algebra: Reasoning with Equations and Inequalities			
A–CED.1★ A–REI.4b	Taking the Square Root of Both Sides	Time to Splash	72
A–SSE.2 A–CED.1★ A–REI.4b	Factoring	On the Shelf	75
A–SSE.2 A–CED.1★ A–REI.4a A–REI.4b	Completing the Square	Curve Ball	79
A–CED.1★ A–REI.4a A–REI.4b	Applying the Quadratic Formula	Divine Proportion	83

(continued)

Standards Correlations

CCSS addressed	Task focus	Task title	Page number
A–SSE.2 A–CED.1★ A–REI.4b	Solving Quadratic Inequalities	Dancing for Charity	86
A–REI.7	Solving Systems Graphically	Constant Acceleration and Constant Speed	90
A–REI.7	Solving Systems Algebraically	The Business of Bicycles	93
Functions: Interpreting Functions			
F–IF.4★	Interpreting Key Features of Quadratic Functions	One-on-One Basketball	97
F–IF.5★	Identifying the Domain of a Quadratic Function	Window Washers	101
F–IF.6★	Identifying the Average Rate of Change	Is the Maximum High Enough?	105
F–IF.7a★	Graphing Quadratic Functions	How High Can a Frog Jump?	108
F–IF.7a★ F–IF.8a	Interpreting Various Forms of Quadratic Functions	Is the Glider Safe?	112
F–IF.7b★	Square Root and Cube Root Functions	Auto Acceleration	116
F–IF.7b★	Absolute Value and Step Functions	Bike Rentals	122
F–IF.7b★	Piecewise Functions	Stop and Go	128
F–IF.8b	Analyzing Exponential Functions	Which Car Is a Better Buy?	134
Functions: Building Functions			
F–BF.1a★	Building Functions from Context	To Drill or Not to Drill?	138
F–BF.1b★	Operating on Functions	Pushing Envelopes	142
F–BF.3	Replacing $f(x)$ with $f(x) + k$ and $f(x + k)$	The Catch	146
F–BF.3	Replacing $f(x)$ with $k \bullet f(x)$ and $f(k \bullet x)$	Fewer Parabolas, Please	150
F–BF.4a	Finding Inverse Functions	Falling Keys	156

(*continued*)

Standards Correlations

CCSS addressed	Task focus	Task title	Page number
Functions: Linear, Quadratic, and Exponential Models			
F–IF.9 F–LE.3★	Comparing Properties of Functions Given in Different Forms	On to Washington!	160
Functions: Trigonometric Functions			
F–TF.8	Proving the Pythagorean Identity	Slick Surfboards	164
Geometry: Congruence			
G–CO.9	Proving the Vertical Angles Theorem	Cutting Kitchen Tiles	168
G–CO.9	Proving Theorems About Angles in Parallel Lines Cut by a Transversal	Retinal Refraction	173
G–CO.10	Proving the Interior Angle Sum Theorem	Sensing Distance	177
G–CO.10	Proving Theorems About Isosceles Triangles	Calibrating Consoles	181
G–CO.10	Proving the Midsegment of a Triangle	Life-Size Support	185
G–CO.10	Proving Centers of Triangles	Sailing Centroid	189
G–CO.11	Proving Properties of Parallelograms	Measuring Bridge Cable Tension	196
G–CO.11	Proving Properties of Special Quadrilaterals	Where Are the Catchers?	201
Geometry: Similarity, Right Triangles, and Trigonometry			
G–SRT.1a	Investigating Properties of Parallelism and the Center	Prettying Up the Pentagon	209
G–SRT.1b	Investigating Scale Factors	The Bigger Picture	213
G–SRT.2	Defining Similarity	Video Game Transformations	218
G–SRT.3	Applying Similarity Using the Angle-Angle (AA) Criterion	True Trusses	223
G–SRT.4	Proving Triangle Similarity Using Side-Angle-Side (SAS) and Side-Side-Side (SSS) Similarity	Down, Down, Down	226
G–SRT.4	Working with Ratio Segments	Suddenly Sinking	230

(*continued*)

Standards Correlations

CCSS addressed	Task focus	Task title	Page number
G–SRT.4	Proving the Pythagorean Theorem Using Similarity	Towering Heights	234
G–SRT.5	Solving Problems Using Similarity and Congruence	Too Tall?	238
G–SRT.6	Defining Trigonometric Ratios	Building a Porch Ramp	243
G–SRT.7	Exploring Sine and Cosine As Complements	School Vegetable Garden	247
G–SRT.8	Calculating Sine, Cosine, and Tangent	Replanting Project	251
G–SRT.8	Calculating Cosecant, Secant, and Cotangent	A Baseball Triangle	255
G–SRT.8	Problem Solving with the Pythagorean Theorem and Trigonometry	Fighting Flames from a Distance	260
Geometry: Circles			
G–C.1 G–C.2	Similar Circles and Central and Inscribed Angles	Following in Archimedes' Footsteps	265
G–C.2	Chord Central Angles Conjecture	Masking the Problem	268
G–C.2	Properties of Tangents of a Circle	The Circus Is in Town! Is It Safe?	271
G–C.3	Constructing Inscribed Circles	First Aid Station	275
G–C.3	Constructing Circumscribed Circles	Building a New Radio Station	279
G–C.3	Proving Properties of Inscribed Quadrilaterals	King Arthur and His Round Table	283
G–C.4 (+)	Constructing Tangent Lines	Designing a T-Shirt Logo	286
G–C.5	Defining Radians	Around the Merry-Go-Round	290
G–C.5	Deriving the Formula for the Area of a Sector	Pizza Specials	294
Geometry: Expressing Geometric Properties with Equations			
G–GPE.1	Deriving the Equation of a Circle	Nurturing an Investment	298

(*continued*)

Standards Correlations

CCSS addressed	Task focus	Task title	Page number
G–GPE.2	Deriving the Equation of a Parabola	A Ball in Flight	303
G–GPE.4	Using Coordinates to Prove Geometric Theorems About Circles and Parabolas	A Circle Graph for Lunch	308
G–GPE.6	Midpoints and Other Points on Line Segments	Points of Shade	312
Geometry: Geometric Measurement and Dimension			
G–GMD.1	Circumference and Area of a Circle	Designing a Tablecloth	317
G–GMD.1 G–GMD.3★	Volumes of Cylinders, Pyramids, Cones, and Spheres	Cylinders of Sand	321
Statistics and Probability: Conditional Probability and the Rules of Probability			
S–CP.1★	Describing Events	Recognizing Equal Sets	325
S–CP.2★	Understanding Independent Events	Tossing Coins	330
S–CP.3★ S–CP.5★ S–CP.6★	Introducing Conditional Probability	Does She Have a Brother or a Sister?	334
S–CP.4★ S–CP.5★ S–CP.6★	Using Two-Way Frequency Tables	How Many in the Survey?	337
S–CP.7★	The Addition Rule	Paying the Bonus?	341
S–CP.8★ (+)	The Multiplication Rule	Playing Cards	345
S–CP.9★ (+)	Combinations and Permutations	On the Court	350
S–CP.9★ (+)	Probability with Combinatorics	The Apple Game	354
Statistics and Probability: Using Probability to Make Decisions			
S–MD.6★ (+)	Making Decisions	Two of a Kind	358
S–MD.7★ (+)	Analyzing Decisions	Changing Channels	362

N–RN.1; N–RN.2 • Number and Quantity

Defining, Rewriting, and Evaluating Rational Exponents

Common Core State Standards

N–RN.1

Explain how the definition of the meaning of rational exponents follows from extending the properties of integer exponents to those values, allowing for a notation for radicals in terms of rational exponents. *For example, we define $5^{1/3}$ to be the cube root of 5 because we want $(5^{1/3})^3 = 5^{(1/3)3}$ to hold, so $(5^{1/3})^3$ must equal 5.*

N–RN.2

Rewrite expressions involving radicals and rational exponents using the properties of exponents.

Problem-Based Task: Population Growth

A town takes a census, or a count of its population, every 10 years. The town uses the census to estimate the population's growth rate. The town's population today is 42,000 people, and the 10-year growth rate is approximately 35%. The town's population can be estimated at any year t using the equation $y = y_0 \bullet (1+r)^{\frac{t}{10}}$, where y_0 is the initial population and r is the 10-year growth rate. What will be the town's approximate population 8 years from today?

Problem-Based Task: Population Growth

Coaching

a. What equation can you use to estimate the population in this situation?

b. What do each of the variables in the equation represent?

c. Which variables can you replace with quantities, and what are the quantities?

d. How can you further simplify the equation before you estimate the population?

e. What will be the town's approximate population 8 years from today?

Problem-Based Task: Population Growth

Coaching Sample Responses

a. What equation can you use to estimate the population in this situation?

$$y = y_0 \bullet (1+r)^{\frac{t}{10}}$$

b. What do each of the variables in the equation represent?

y is the population in 8 years, y_0 is the population today, r is the 10-year growth rate, and t is the time in years.

c. Which variables can you replace with quantities, and what are the quantities?

y_0, r, and t are all given. y_0 is 42,000, r is 0.35, and t is 8.

d. How can you further simplify the equation before you estimate the population?

Write the original equation with quantities in place of variables.

$$y = 42{,}000 \bullet (1+0.35)^{\frac{8}{10}}$$

The rational exponent can be reduced to simplest form. Note that the original root is even; therefore, the solution must be positive. Because all values in the equation are positive, no absolute value symbols are needed.

$$y = 42{,}000 \bullet (1+0.35)^{\frac{8}{10}}$$

$$y = 42{,}000 \bullet (1+0.35)^{\frac{4}{5}}$$

e. What will be the town's approximate population 8 years from today?

Use a calculator to find the town's approximate population 8 years from today.

$$y = 42{,}000 \bullet (1+0.35)^{\frac{4}{5}}$$

$$y = 42{,}000 \bullet (1.35)^{\frac{4}{5}}$$

$$y = 42{,}000 \bullet (1.0271)$$

$$y \approx 53{,}396.9$$

The town's approximate population will be 53,397 people 8 years from today.

N–RN.2; N–RN.3 • Number and Quantity

Rational and Irrational Numbers and Their Properties

Common Core State Standards

N–RN.2

Rewrite expressions involving radicals and rational exponents using the properties of exponents.

N–RN.3

Explain why the sum or product of two rational numbers is rational; that the sum of a rational number and an irrational number is irrational; and that the product of a nonzero rational number and an irrational number is irrational.

Problem-Based Task: Estimating Depreciation

Yasmina is buying a new car. To estimate how her car will decrease in value, or depreciate, she looks at the price of older versions of the same car. She finds a similar car that is 2.5 years old. The original price of the car was \$22,000, and the current selling price is \$16,905. She knows the equation $c = 22{,}000 \bullet d^t$ can be used to estimate the value of the car in any year t after being purchased for \$22,000. The value d is used to calculate the new value of the car each year. Write an equation to help Yasmina estimate the value of her new car, c, in any year t, if the original purchase price is \$22,000.

Problem-Based Task: Estimating Depreciation

Coaching

a. Which variable needs to be determined in the equation $c = 22{,}000 \bullet d^t$ so that Yasmina can estimate the value of a car in year t?

b. Use the older car to help Yasmina complete the equation. Which variables are known for this car?

c. Substitute the known quantities for the variables in the equation.

d. Use inverse operations to find an equivalent equation in the form $x^a = b$, where x is the only variable.

e. Solve for the variable needed to complete the equation.

f. What is the equation Yasmina can use to estimate the value of a car purchased for \$22,000 in any year t?

g. What will be the value of her new car 6 years after she has purchased it?

Problem-Based Task: Estimating Depreciation

Coaching Sample Responses

a. Which variable needs to be determined in the equation $c = 22{,}000 \bullet d^t$ so that Yasmina can estimate the value of a car in year t?

The variable c represents the value of the car in year t; t is the independent variable and c is the dependent variable. The only variable that needs to be determined is d.

b. Use the older car to help Yasmina complete the equation. Which variables are known for this car?

For the older car, Yasmina knows the current value, c, and the time, t, in years.

c. Substitute the known quantities for the variables in the equation.

$c = \$16{,}905$

$t = 2.5$ years

$16{,}905 = 22{,}000 \bullet d^{2.5}$

d. Use inverse operations to find an equivalent equation in the form $x^a = b$, where x is the only variable.

$16{,}905 = 22{,}000 \bullet d^{2.5}$

$0.768 \approx d^{2.5}$

e. Solve for the variable needed to complete the equation.

$0.768 \approx d^{2.5}$

$0.768 \approx d^{2.5}$

$0.768 \approx d^{\frac{5}{2}}$

$(0.768)^{\frac{2}{5}} \approx \left(d^{\frac{5}{2}}\right)^{\frac{2}{5}}$

$0.90 \approx d$

f. What is the equation Yasmina can use to estimate the value of a car purchased for $22,000 in any year *t*?

Replace *d* with the calculated value of 0.90.

$c = 22{,}000(0.90)^t$

g. What will be the value of her new car 6 years after she has purchased it?

Replace *t* with 6 in the equation found in part f.

$c = 22{,}000(0.90)^6$

$c = 11{,}691.70$

After 6 years, the car will be worth approximately $11,692, or about half of its original value.

N–CN.1 • Number and Quantity

Defining Complex Numbers, i, and i^2

Common Core State Standard

N–CN.1

Know there is a complex number i such that $i^2 = -1$, and every complex number has the form $a + bi$ with a and b real.

Problem-Based Task: Representing Impedance

Impedance is the measure of an object's resistance to an electric current, or its opposition to the flow of a current. Complex numbers are used to represent the impedance of an element in a circuit. A certain element has a voltage of 21 volts and a current of 1.25 milliamperes. If impedance is equal to $V + Ii$, where I is in milliamperes, then what is the element's impedance?

Problem-Based Task: Representing Impedance

Coaching

a. Which quantity will be the real part of the complex number?

b. Which quantity will be the multiple of i in the complex number?

c. How is the multiple of i used to write the imaginary part of a complex number?

d. How are the real and imaginary parts used to write a complex number?

e. What is the element's impedance?

Problem-Based Task: Representing Impedance

Coaching Sample Responses

a. Which quantity will be the real part of the complex number?

The voltage, 21, is the real part of the complex number.

b. Which quantity will be the multiple of i in the complex number?

The current, 1.25, is the multiple of i in the complex number.

c. How is the multiple of i used to write the imaginary part of a complex number?

The product of the multiple of i and i is the imaginary part: $1.25 \bullet i = 1.25i$.

d. How are the real and imaginary parts used to write a complex number?

The sum of the real and imaginary parts is a complex number.

$21 + 1.25i$

e. What is the element's impedance?

The element's impedance is $21 + 1.25i$.

N–CN.2 • Number and Quantity

Adding and Subtracting Complex Numbers

Common Core State Standard

N–CN.2

Use the relation $i^2 = -1$ and the commutative, associative, and distributive properties to add, subtract, and multiply complex numbers.

Problem-Based Task: Elements in Series in a Circuit

The impedance of an element can be represented using the complex number, $V + Ii$, where V is the element's voltage and I is the element's current in milliamperes. If two elements are in a circuit in series, the total impedance is the sum of the impedance of each element. The following diagram of a circuit contains two elements, 1 and 2, in series.

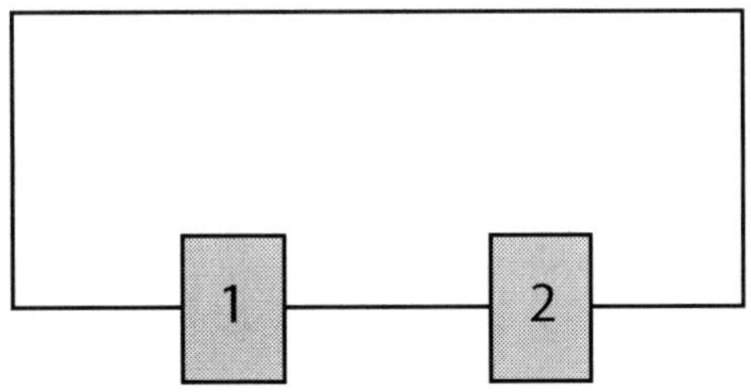

Element 1 has a voltage of 30.5 volts and a current of 2.8 milliamperes. Element 2 has a voltage of 19 volts and a current of 3 milliamperes. What is the total impedance of the circuit?

Problem-Based Task: Elements in Series in a Circuit

Coaching

a. How can the voltage and current be used to write a complex number to represent the impedance of each element?

b. What is the impedance of each element?

c. What operation is used to find the total impedance if the elements are in series?

d. Describe how to find the sum or difference of two complex numbers by using their real and imaginary parts.

e. What is the total impedance of the circuit?

Problem-Based Task: Elements in Series in a Circuit

Coaching Sample Responses

a. How can the voltage and current be used to write a complex number to represent the impedance of each element?

The impedance of each element can be written in the form $V + Ii$, where V is the voltage and I is the current in milliamperes.

b. What is the impedance of each element?

Element 1: $30.5 + 2.8i$

Element 2: $19 + 3i$

c. What operation is used to find the total impedance if the elements are in series?

Addition is used to find total impedance.

Therefore, find the sum of the impedance of each element.

d. Describe how to find the sum or difference of two complex numbers by using their real and imaginary parts.

To find the sum of two complex numbers, find the sum of the real parts, and find the multiple of i by summing the coefficients of i from each complex number.

To find the difference of two complex numbers, find the difference of the real parts, and find the multiple of i by subtracting the coefficients of i from each complex number.

e. What is the total impedance of the circuit?

$(30.5 + 2.8i) + (19 + 3i)$

$= (30.5 + 19) + (2.8 + 3)i$

$= 49.5 + 5.8i$

The total impedance of the circuit is $49.5 + 5.8i$.

N–CN.2 • Number and Quantity

Multiplying Complex Numbers

Common Core State Standard

N–CN.2

Use the relation $i^2 = -1$ and the commutative, associative, and distributive properties to add, subtract, and multiply complex numbers.

Problem-Based Task: Elements in Parallel in a Circuit

The impedance of an element can be represented using the complex number, $V + Ii$, where V is the element's voltage and I is the element's current in milliamperes. The following diagram of a circuit contains two elements, 1 and 2, in parallel.

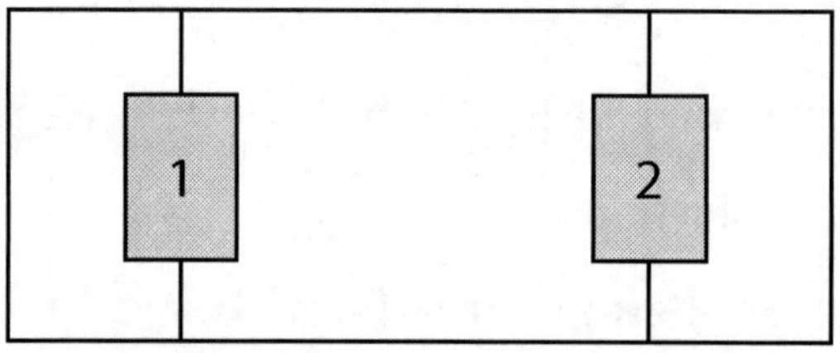

If the impedance of element 1 is $Z_1 = 15 + i$, and the impedance of element 2 is $Z_2 = 10 + 2i$, the total impedance of the two elements in parallel is $\frac{1}{Z_1} + \frac{1}{Z_2}$. What is the total impedance for the two elements in parallel? Leave your response as a fraction.

Problem-Based Task: Elements in Parallel in a Circuit

Coaching

a. What is the reciprocal of each impedance?

b. How can a common denominator be calculated?

c. Find the common denominator.

d. Use the factor needed to calculate the common denominator to write an equivalent reciprocal for each impedance.

e. What is the sum of these reciprocals, written as a fraction?

f. What is the total impedance for the two elements in parallel? Leave your response as a fraction.

Problem-Based Task: Elements in Parallel in a Circuit

Coaching Sample Responses

a. What is the reciprocal of each impedance?

$$\frac{1}{Z_1}=\frac{1}{15+i} \qquad \frac{1}{Z_2}=\frac{1}{10+2i}$$

b. How can a common denominator be calculated?

The product of the two existing denominators can be used as the common denominator.

c. Find the common denominator.

Multiply both terms in the first polynomial by both terms in the second polynomial. Find the product of the first terms, outside terms, inside terms, and last terms.

$(15 + i)(10 + 2i)$

$= 150 + 30i + 10i + 2i^2$

$= 148 + 40i$

d. Use the factor needed to calculate the common denominator to write an equivalent reciprocal for each impedance.

$$\frac{1}{Z_1}=\frac{1}{15+i}=\frac{1}{15+i}\bullet\frac{10+2i}{10+2i}=\frac{10+2i}{148+40i}$$

$$\frac{1}{Z_2}=\frac{1}{10+2i}=\frac{1}{10+2i}\bullet\frac{15+i}{15+i}=\frac{15+i}{148+40i}$$

e. What is the sum of these reciprocals, written as a fraction?

$$\frac{10+2i}{148+40i}+\frac{15+i}{148+40i}=\frac{25+3i}{148+40i}$$

f. What is the total impedance for the two elements in parallel? Leave your response as a fraction.

The total impedance for the two elements in parallel is $\frac{25+3i}{148+40i}$.

Common Core State Standards

N–CN.7

Solve quadratic equations with real coefficients that have complex solutions.

N–CN.9

(+) Know the Fundamental Theorem of Algebra; show that it is true for quadratic polynomials.

Problem-Based Task: All Roads Lead to the Same Destination

Solve $3x^2 + 5 = 0$ by the following three methods. Show that all three methods give the same solutions.

- Use the quadratic formula.
- Use this property of square roots: If $x^2 = a$, then $x = \pm\sqrt{a}$.
- Show through factoring that $3x^2 + 5$ can be written as a sum of two squares.

Problem-Based Task: All Roads Lead to the Same Destination

Coaching

a. The quadratic formula states that if $ax^2 + bx + c = 0$, then $x = \frac{-b \pm \sqrt{b^2 - 4ac}}{2a}$. What are the values of a, b, and c in the equation $3x^2 + 5 = 0$?

b. How can you solve the equation with the quadratic formula?

c. Recall that if $x^2 = a$, then $x = \pm\sqrt{a}$. How can you solve the equation by using this property of square roots? How can you show that your solutions by this method match the solutions in part b?

d. How can you solve the equation by factoring? (*Hint*: Write the given equation. Then write the equation again, showing that $3x^2 + 5$ is the sum of two squares.)

e. How can you apply this identity that shows how to factor the sum of two squares: $a^2 + b^2 = (a + bi)(a - bi)$?

f. The Zero Product Property allows you to state that if $(a + bi)(a - bi) = 0$, then either $a + bi = 0$ or $a - bi = 0$. How can you use the Zero Product Property to finish solving and then write your solutions in the form that matches the solutions in parts b and c?

Problem-Based Task: All Roads Lead to the Same Destination

Coaching Sample Responses

a. The quadratic formula states that if $ax^2 + bx + c = 0$, then $x=\dfrac{-b\pm\sqrt{b^2-4ac}}{2a}$. What are the values of a, b, and c in the equation $3x^2 + 5 = 0$?

Since the equation is already written in standard form, $a = 3$, $b = 0$, and $c = 5$.

b. How can you solve the equation by the quadratic formula?

$3x^2 + 5 = 0$	Write the given equation.
$x=\dfrac{-0\pm\sqrt{0^2-4(3)(5)}}{2(3)}$	Substitute values for a, b, and c: $a = 3$, $b = 0$, $c = 5$.
$x=\dfrac{\pm\sqrt{-60}}{6}$	Simplify.
$x=\dfrac{\pm i\sqrt{60}}{6}$	For any positive real number a, $\sqrt{-a}=i\sqrt{a}$.
$x=\dfrac{\pm 2i\sqrt{15}}{6}$	Simplify.
$x=\pm\left(\dfrac{\sqrt{15}}{3}\right)i$	

c. Recall that if $x^2 = a$, then $x=\pm\sqrt{a}$. How can you solve the equation by using this property of square roots? How can you show that your solutions by this method match the solutions in part b?

$3x^2 + 5 = 0$	Write the given equation.
$3x^2 = -5$	Subtract 5 from both sides.
$x^2=-\dfrac{5}{3}$	Divide both sides by 3.
$x=\pm\sqrt{-\dfrac{5}{3}}$	Apply this property of square roots: If $x^2 = a$, then $x=\pm\sqrt{a}$.

$x=\pm i\sqrt{\frac{5}{3}}$	For any positive real number a, $\sqrt{-a}=i\sqrt{a}$.
$x=\pm i\frac{\sqrt{5}}{\sqrt{3}}$	For any real numbers a and b with $b\neq 0$, $\sqrt{\frac{a}{b}}=\frac{\sqrt{a}}{\sqrt{b}}$.
$x=\pm i\frac{\sqrt{5}}{\sqrt{3}}\bullet\frac{\sqrt{3}}{\sqrt{3}}$	Multiply by $\frac{\sqrt{3}}{\sqrt{3}}$, which is equal to 1.
$x=\pm i\frac{\sqrt{5\bullet 3}}{\sqrt{3\bullet 3}}$	For any real numbers a and b, $\sqrt{a}\bullet\sqrt{b}=\sqrt{ab}$.
$x=\pm i\frac{\sqrt{15}}{\sqrt{9}}$	Simplify.
$x=\pm i\frac{\sqrt{15}}{3}$	
$x=\pm\left(\frac{\sqrt{15}}{3}\right)i$	

d. How can you solve the equation by factoring? (*Hint*: Write the given equation. Then write the equation again, showing that $3x^2 + 5$ is the sum of two squares.)

$3x^2 + 5 = 0$	Write the given equation.
$\left(x\sqrt{3}\right)^2+\left(\sqrt{5}\right)^2=0$	Show that $3x^2 + 5$ is the sum of two squares.

e. How can you apply this identity that shows how to factor the sum of two squares: $a^2 + b^2 = (a + bi)(a - bi)$?

$3x^2 + 5 = 0$	Given equation
$\left(x\sqrt{3}\right)^2+\left(\sqrt{5}\right)^2=0$	Show that $3x^2 + 5$ is the sum of two squares.
$\left(x\sqrt{3}+i\sqrt{5}\right)\left(x\sqrt{3}-i\sqrt{5}\right)=0$	Apply the identity $a^2 + b^2 = (a + bi)(a - bi)$.

f. The Zero Product Property allows you to state that if $(a + bi)(a - bi) = 0$, then either $a + bi = 0$ or $a - bi = 0$. How can you use the Zero Product Property to finish solving and then write your solutions in the form that matches the solutions in parts b and c?

Start with the factorization found in part e.

$$\left(x\sqrt{3}+i\sqrt{5}\right)\left(x\sqrt{3}-i\sqrt{5}\right)=0 \qquad a^2 + b^2 = (a + bi)(a - bi)$$

$$x\sqrt{3}+i\sqrt{5}=0 \text{ or } x\sqrt{3}-i\sqrt{5}=0$$ If $(a + bi)(a - bi) = 0$, then either $a + bi = 0$ or $a - bi = 0$.

$$x\sqrt{3}=-i\sqrt{5} \text{ or } x\sqrt{3}=i\sqrt{5}$$

$$x=-\frac{\sqrt{5}}{\sqrt{3}}i \text{ or } x=\frac{\sqrt{5}}{\sqrt{3}}i$$

$$x=-\frac{\sqrt{5}}{\sqrt{3}}\bullet\frac{\sqrt{3}}{\sqrt{3}}i \text{ or } x=\frac{\sqrt{5}}{\sqrt{3}}\bullet\frac{\sqrt{3}}{\sqrt{3}}i$$

$$x=-\left(\frac{\sqrt{15}}{3}\right)i \text{ or } x=\left(\frac{\sqrt{15}}{3}\right)i$$

$$x=\pm\left(\frac{\sqrt{15}}{3}\right)i$$

Common Core State Standard

N–CN.8

(+) Extend polynomial identities to the complex numbers. *For example, rewrite* $x^2 + 4$ *as* $(x + 2i)(x - 2i)$.

Problem-Based Task: Making a Complex Expression Simple

Kiana is working on a homework problem and has to simplify the expression $\left(\frac{1}{2}+\frac{\sqrt{3}}{2}i\right)^3$. She must then complete the following equation and explain why it is an identity: $\left(\frac{1}{2}+\frac{\sqrt{3}}{2}i\right)^3 \bullet a = \_\_$.

Problem-Based Task: Making a Complex Expression Simple

Coaching

a. How can Kiana simplify $\left(\frac{1}{2}+\frac{\sqrt{3}}{2}i\right)^2$ by using the Distributive Property?

b. How can Kiana use her result from part a along with the fact that $\left(\frac{1}{2}+\frac{\sqrt{3}}{2}i\right)^3=\left(\frac{1}{2}+\frac{\sqrt{3}}{2}i\right)\bullet\left(\frac{1}{2}+\frac{\sqrt{3}}{2}i\right)^2$ to simplify $\left(\frac{1}{2}+\frac{\sqrt{3}}{2}i\right)^3$?

c. The correct answer to part b is the value of $\left(\frac{1}{2}+\frac{\sqrt{3}}{2}i\right)^3$. How can Kiana use this value to complete the equation $\left(\frac{1}{2}+\frac{\sqrt{3}}{2}i\right)^3\bullet a=\_\_$?

d. The correct completed equation in part c is an identity. Why is it an identity?

Problem-Based Task: Making a Complex Expression Simple

Coaching Sample Responses

a. How can Kiana simplify $\left(\frac{1}{2}+\frac{\sqrt{3}}{2}i\right)^2$ by using the Distributive Property?

Factor the expression, and then distribute the first factor over the second factor.

$$\left(\frac{1}{2}+\frac{\sqrt{3}}{2}i\right)^2=\left(\frac{1}{2}+\frac{\sqrt{3}}{2}i\right)\left(\frac{1}{2}+\frac{\sqrt{3}}{2}i\right)$$

$$\left(\frac{1}{2}+\frac{\sqrt{3}}{2}i\right)^2=\frac{1}{2}\bullet\frac{1}{2}+\frac{1}{2}\bullet\frac{\sqrt{3}}{2}i+\frac{1}{2}\bullet\frac{\sqrt{3}}{2}i+\frac{\sqrt{3}}{2}i\bullet\frac{\sqrt{3}}{2}i$$

$$\left(\frac{1}{2}+\frac{\sqrt{3}}{2}i\right)^2=\frac{1}{4}+\frac{\sqrt{3}}{4}i+\frac{\sqrt{3}}{4}i+\frac{3}{4}i^2$$

$$\left(\frac{1}{2}+\frac{\sqrt{3}}{2}i\right)^2=\frac{1}{4}+\frac{2\sqrt{3}}{4}i+\frac{3}{4}i^2$$

$$\left(\frac{1}{2}+\frac{\sqrt{3}}{2}i\right)^2=\frac{1}{4}+\frac{\sqrt{3}}{2}i+\frac{3}{4}i^2$$

$$\left(\frac{1}{2}+\frac{\sqrt{3}}{2}i\right)^2=\frac{1}{4}+\frac{\sqrt{3}}{2}i+\frac{3}{4}(-1)$$

$$\left(\frac{1}{2}+\frac{\sqrt{3}}{2}i\right)^2=\frac{1}{4}+\frac{\sqrt{3}}{2}i-\frac{3}{4}$$

$$\left(\frac{1}{2}+\frac{\sqrt{3}}{2}i\right)^2=-\frac{1}{2}+\frac{\sqrt{3}}{2}i$$

b. How can Kiana use her result from part a along with the fact that $\left(\frac{1}{2}+\frac{\sqrt{3}}{2}i\right)^3=\left(\frac{1}{2}+\frac{\sqrt{3}}{2}i\right)\bullet\left(\frac{1}{2}+\frac{\sqrt{3}}{2}i\right)^2$ to simplify $\left(\frac{1}{2}+\frac{\sqrt{3}}{2}i\right)^3$?

Use the properties of exponents to factor the expression, and then substitute the result from part a for $\left(\frac{1}{2}+\frac{\sqrt{3}}{2}i\right)^2$.

$$\left(\frac{1}{2}+\frac{\sqrt{3}}{2}i\right)^3=\left(\frac{1}{2}+\frac{\sqrt{3}}{2}i\right)\bullet\left(\frac{1}{2}+\frac{\sqrt{3}}{2}i\right)^2$$

$$\left(\frac{1}{2}+\frac{\sqrt{3}}{2}i\right)^3=\left(\frac{1}{2}+\frac{\sqrt{3}}{2}i\right)\bullet\left(-\frac{1}{2}+\frac{\sqrt{3}}{2}i\right)$$

$$\left(\frac{1}{2}+\frac{\sqrt{3}}{2}i\right)^3=\frac{1}{2}\bullet\left(-\frac{1}{2}\right)+\frac{1}{2}\bullet\frac{\sqrt{3}}{2}i+\left(-\frac{1}{2}\right)\bullet\frac{\sqrt{3}}{2}i+\frac{\sqrt{3}}{2}i\bullet\frac{\sqrt{3}}{2}i$$

$$\left(\frac{1}{2}+\frac{\sqrt{3}}{2}i\right)^3=-\frac{1}{4}+\frac{\sqrt{3}}{4}i-\frac{\sqrt{3}}{4}i+\frac{3}{4}i^2$$

$$\left(\frac{1}{2}+\frac{\sqrt{3}}{2}i\right)^3=-\frac{1}{4}+\frac{3}{4}i^2$$

$$\left(\frac{1}{2}+\frac{\sqrt{3}}{2}i\right)^3=-\frac{1}{4}+\frac{3}{4}(-1)$$

$$\left(\frac{1}{2}+\frac{\sqrt{3}}{2}i\right)^3=-\frac{1}{4}-\frac{3}{4}$$

$$\left(\frac{1}{2}+\frac{\sqrt{3}}{2}i\right)^3=-1$$

c. The correct answer to part b is the value of $\left(\frac{1}{2}+\frac{\sqrt{3}}{2}i\right)^3$. How can Kiana use this value to complete the equation: $\left(\frac{1}{2}+\frac{\sqrt{3}}{2}i\right)^3 \bullet a = \_\_$?

$\left(\frac{1}{2}+\frac{\sqrt{3}}{2}i\right)^3 \bullet a = -1 \bullet a = -a$, so the completed equation is $\left(\frac{1}{2}+\frac{\sqrt{3}}{2}i\right)^3 \bullet a = -a$.

d. The correct completed equation in part c is an identity. Why is it an identity?

The equation $\left(\frac{1}{2}+\frac{\sqrt{3}}{2}i\right)^3 \bullet a = -a$ is an identity because it is true for all values of a.

NAME:

A–SSE.1a★ • Algebra

Identifying Terms, Factors, and Coefficients

Common Core State Standard

A–SSE.1

Interpret expressions that represent a quantity in terms of its context.★

a. Interpret parts of an expression, such as terms, factors, and coefficients.

Problem-Based Task: Deck the Deck

Shanna wants to decorate the triangular deck behind her house. The base of the triangle is 10 meters shorter than the altitude. What are the terms, factors, and coefficients of the quadratic expression that represents the area of the deck to be decorated?

Problem-Based Task: Deck the Deck

Coaching

a. How can you represent the altitude of the triangle?

b. What algebraic expression represents the base of the triangle in terms of the altitude?

c. What algebraic expression represents the area of the triangle?

d. What units will be used to measure the area of the deck?

e. Simplify your expression for the area using the Distributive Property. Is this a quadratic expression?

f. How many terms does the expression from part e include?

g. What are the factors that make up each of the terms?

h. What are the coefficients of each term?

Problem-Based Task: Deck the Deck

Coaching Sample Responses

a. How can you represent the altitude of the triangle?

Let a represent the altitude of the triangle.

b. What algebraic expression represents the base of the triangle in terms of the altitude?

Since the base of the triangle is 10 meters shorter than the altitude, it can be represented by the expression $a - 10$.

c. What algebraic expression represents the area of the triangle?

The formula for the area of a triangle is $\frac{1}{2}bh$, where b represents the base of the triangle and h represents the altitude (or height) of the triangle.

Substitute the expressions from parts a and b to express the area of this triangle as $\frac{1}{2}a(a-10)$.

d. What units will be used to measure the area of the deck?

Square units are used to measure area, so the deck will be measured in square meters.

e. Simplify your expression for the area using the Distributive Property. Is this a quadratic expression?

After simplifying, the expression becomes $\frac{1}{2}a^2-5a$. This is a quadratic expression with $c = 0$.

f. How many terms does the expression from part e include?

There are two terms in the expression $\frac{1}{2}a^2-5a$. The terms are $\frac{1}{2}a^2$ and $-5a$.

g. What are the factors that make up each of the terms?

The factors are the quantities whose products are terms. The product of $\frac{1}{2}a^2$ is made up of the factors $\frac{1}{2}$ and a^2. The product of $-5a$ is made up of the factors -5 and a.

h. What are the coefficients of each term?

The coefficient of the term $\frac{1}{2}a^2$ is $\frac{1}{2}$. The coefficient of the term $-5a$ is -5.

A–SSE.1b★ • Algebra

Interpreting Complicated Expressions

Common Core State Standard

A–SSE.1

Interpret expressions that represent a quantity in terms of its context.★

b. Interpret complicated expressions by viewing one or more of their parts as a single entity. *For example, interpret $P(1 + r)^n$ as the product of P and a factor not depending on P.*

Problem-Based Task: Puppy Pen

Oscar has 20 feet of fencing. He wants to build a rectangular pen for his new puppy. Give a formula for the area of the puppy's pen and show that the formula is quadratic. If 2 feet of fencing are damaged and cannot be used, how does this affect the pen's area?

Problem-Based Task: Puppy Pen

Coaching

a. Let l be the length of the pen and w be the width of the pen. Write an expression for the perimeter of the rectangular pen.

b. Use what you know about the perimeter of the pen to write an expression for the pen's width in terms of the pen's length, l.

c. How can the area of the pen be expressed in terms of l?

d. Show that the area formula is quadratic.

e. If 2 feet of fencing are damaged, what is the pen's perimeter?

f. How does the change in perimeter affect the expression for width?

g. How will the area change if 2 feet of fencing are damaged?

Problem-Based Task: Puppy Pen

Coaching Sample Responses

a. Let l be the length of the pen and w be the width of the pen. Write an expression for the perimeter of the rectangular pen.

The perimeter of the pen can be expressed as $2l + 2w$ feet.

b. Use what you know about the perimeter of the pen to write an expression for the pen's width in terms of the pen's length, l.

We know that the perimeter is 20 feet. Set the perimeter equal to 20 and solve for w.

$2l + 2w = 20$

$2w = 20 - 2l$

$w = 10 - l$ feet

c. How can the area of the pen be expressed in terms of l?

The area of a rectangle is found by multiplying the length by the width.

$A = l \bullet w = l(10 - l)$ square feet

d. Show that the area formula is quadratic.

Multiply the factors of the expression.

$A = l(10 - l) = 10l - l^2$ square feet

This is a quadratic equation, with $a = -1$, $b = 10$, and $c = 0$.

e. If 2 feet of fencing are damaged, what is the pen's perimeter?

If 2 feet of fencing are damaged, the pen's perimeter will decrease from 20 feet to 18 feet.

f. How does the change in perimeter affect the expression for width?

Substitute 18 for the perimeter to find the new expression for width.

$2l + 2w = 18$

$2w = 18 - 2l$

$w = 9 - l$

g. How will the area change if 2 feet of fencing are damaged?

Substitute the new expression for width into the area formula and multiply the factors.

$A = l(9 - l) = 9l - l^2$ square feet

If 2 feet of fencing are damaged, the area will change from $10l - l^2$ square feet to $9l - l^2$ square feet. The area decreases by the length of the pen.

Common Core State Standards

A–CED.2

Create equations in two or more variables to represent relationships between quantities; graph equations on coordinate axes with labels and scales.★

A–SSE.3

Choose and produce an equivalent form of an expression to reveal and explain properties of the quantity represented by the expression.★

a. Factor a quadratic expression to reveal the zeros of the function it defines.

Problem-Based Task: Parabolic Party Streamers

Students in a physics class are testing the pull of gravity at varying heights. With permission of the building manager, each student went to a different floor of a tall office building and tossed a roll of paper streamer up into the air from the window. Another student videotaped the streamers' paths downward so that the class could determine the approximate equations of the parabolas the streamers created as they unraveled. When a streamer was thrown from the highest story of the building, the students determined that the distance, in inches, between the streamer and the ground t seconds after the streamer was thrown could be expressed by $h(t) = -16t^2 + 32t + 673$. After how many seconds and at what height was the streamer at its maximum distance from the ground?

Problem-Based Task: Parabolic Party Streamers

Coaching

a. What is the formula stated in the problem?

b. Is the formula written in standard form?

c. What are the values of *a*, *b*, and *c*?

d. How do you determine when the streamer reached its highest point?

e. After how many seconds was the streamer at its maximum height?

f. What height did the streamer reach at its maximum?

Problem-Based Task: Parabolic Party Streamers

Coaching Sample Responses

a. What is the formula stated in the problem?

$h(t) = -16t^2 + 32t + 673$

b. Is the formula written in standard form?

Yes, the equation $h(t) = -16t^2 + 32t + 673$ is written in the form $f(x) = ax^2 + bx + c$, which is the standard form of a quadratic equation.

c. What are the values of a, b, and c?

The value of a is –16.

The value of b is 32.

The value of c is 673.

d. How do you determine when the streamer reached its highest point?

Identify the y-coordinate of the vertex by using the formula for axis of symmetry to find the input value, t. Then substitute that into the function to find the output value, $h(t)$.

e. After how many seconds was the streamer at its maximum height?

$$x = \frac{-b}{2a}$$

$$x = \frac{-(32)}{2(-16)}$$

$$x = \frac{-32}{-32}$$

$$x = 1$$

The streamer was at its maximum height after 1 second.

f. What height did the streamer reach at its maximum?

$h(t) = -16t^2 + 32t + 673$

$h(1) = -16(1)^2 + 32(1) + 673$

$h(1) = 689$

The streamer reached its maximum height at 689 inches.

NAME:

A–CED.2★; A–SSE.3a★ • Algebra

Creating and Graphing Equations Using the *x*-intercepts

Common Core State Standards

A–CED.2

Create equations in two or more variables to represent relationships between quantities; graph equations on coordinate axes with labels and scales.★

A–SSE.3

Choose and produce an equivalent form of an expression to reveal and explain properties of the quantity represented by the expression.★

a. Factor a quadratic expression to reveal the zeros of the function it defines.

Problem-Based Task: It's Show Time!

The school's drama department is putting on a production. Instead of using the school's indoor stage, the department head decided to build an outdoor stage to accommodate a greater audience. The width of the outdoor stage is 6 feet less than its length. If the total area of the outdoor stage is 720 ft^2, what are the length and width of the outdoor stage?

Problem-Based Task: It's Show Time!

Coaching

a. What are the unknowns?

b. What would a diagram of this situation look like?

c. What formula do you need to use to find the length and width of the outdoor stage?

d. How can you use the unknowns and the formula from part c to begin to determine the length and width of the outdoor stage?

e. How could you begin to solve for length and width?

f. What is the best method to find the length and width of the outdoor stage?

g. Are all values of x appropriate for this context?

h. What are the length and width of the outdoor stage?

Problem-Based Task: It's Show Time!

Coaching Sample Responses

a. What are the unknowns?

The unknowns are the length and width of the outdoor stage.

Length is represented by x and width is represented by $x - 6$.

b. What would a diagram of this situation look like?

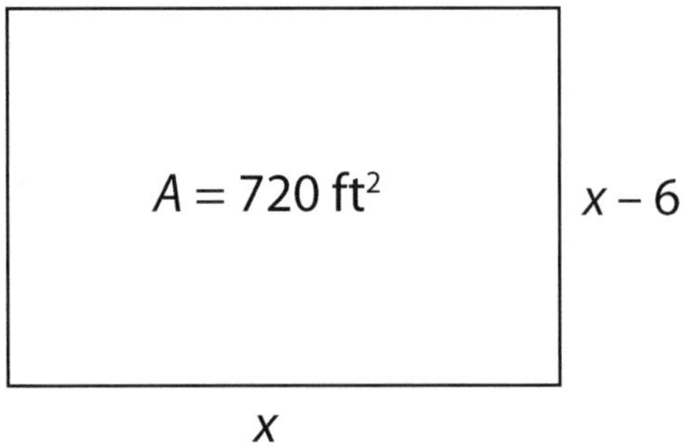

c. What formula do you need to find the length and width of the outdoor stage?

You need to use the formula for the area of a rectangle: $A = lw$.

d. How can you use the unknowns and the formula from part c to begin to determine the length and width of the outdoor stage?

Substitute the equations representing the unknowns into the area formula.

$A = lw$

$720 = (x)(x - 6)$

e. How could you begin to solve for length and width?

Rearrange the equation and set it equal to 0.

$720 = (x)(x - 6)$

$x(x - 6) - 720 = 0$

f. What is the best method to find the length and width of the outdoor stage?

The equation $x(x - 6) - 720 = 0$ can be easily factored; therefore, the best method for finding the length and width is factoring.

$x(x - 6) - 720 = 0$

$x^2 - 6x - 720 = 0$

$(x - 30)(x + 24) = 0$

$x - 30 = 0$ and $x + 24 = 0$

$x = 30$ and $x = -24$

g. Are all values of x appropriate for this context?

Determine the length and width using each x-value.

The length $= x$ and the width $= x - 6$.

When $x = -24$, the length is -24 and the width is $-24 - 6 = -30$.

Substituting -24 results in both a negative length and a negative width. Length and width cannot be negative, so $x = -24$ is not appropriate to the context.

When $x = 30$, the length is 30 and the width is $30 - 6 = 24$.

Substituting 30 for x results in positive values for both the length and the width.

The only value of x that is appropriate for this context is 30 feet.

h. What are the length and width of the outdoor stage?

The length of the outdoor stage is 30 feet.

The width of the outdoor stage is 6 feet less than the length and thus is 24 feet.

NAME:

A–CED.2★; A–SSE.3b★ • Algebra

Creating and Graphing Equations Using Vertex Form

Common Core State Standards

A–CED.2

Create equations in two or more variables to represent relationships between quantities; graph equations on coordinate axes with labels and scales.★

A–SSE.3

Choose and produce an equivalent form of an expression to reveal and explain properties of the quantity represented by the expression.★

b. Complete the square in a quadratic expression to reveal the maximum or minimum value of the function it defines.

Problem-Based Task: Rocket Heights

The path of a model rocket launched from the ground is parabolic. Cruz built a reusable model rocket using a kit. After launching it a few times from level ground, Cruz determined that the rocket could reach a maximum height of 400 feet after traveling a horizontal distance of 15 feet. On the flight where it reached the maximum height, the rocket landed 30 feet away from the launch site. Cruz's friends asked him to launch the rocket again in an area with several trees. He's worried that the rocket will be destroyed if the trees block its path—especially since a 115-foot-tall pine tree stands just 2 feet from the landing spot. Will Cruz's rocket make it over the pine tree?

Problem-Based Task: Rocket Heights

Coaching

a. For which quantities do you have a value?

b. Substitute the known values into the vertex form of a quadratic equation.

c. What is still unknown?

d. How could you solve for this?

e. What is the quadratic equation that would allow Cruz to calculate the height of the rocket for any distance?

f. Will Cruz's rocket make it over the pine tree?

Problem-Based Task: Rocket Heights

Coaching Sample Responses

a. For which quantities do you have a value?

The maximum height of the rocket is 400 feet at a horizontal distance of 15 feet.

This information is the vertex, (15, 400).

The rocket landed 30 feet from the launch location.

This information is one x-intercept, (30, 0).

The second x-intercept is (0, 0), the initial launch location.

b. Substitute the known values into the vertex form of a quadratic equation.

Substitute (15, 400) for (h, k) and (30, 0) for (x, y).

$$y = a(x - h)^2 + k$$

$$0 = a(30 - 15)^2 + 400$$

c. What is still unknown?

By looking at the equation in part b, it can be seen that the value of a is still unknown.

d. How could you use solve for this?

Use algebra to solve for the unknown variable.

$$0 = a(30 - 15)^2 + 400$$

$$-400 = a(30 - 15)^2$$

$$-400 = a(15)^2$$

$$-400 = 225a$$

$$a = -\frac{16}{9}$$

e. What is the quadratic equation that would allow Cruz to calculate the height of the rocket for any distance?

Substitute the value of a, $-\frac{16}{9}$, and the vertex, (15, 400), into the vertex form, $f(x) = a(x - h)^2 + k$.

$$f(x)=-\frac{16}{9}(x-15)^2+400$$

The quadratic equation that would allow Cruz to calculate the height of the rocket is

$$f(x)=-\frac{16}{9}(x-15)^2+400.$$

f. Will Cruz's rocket make it over the pine tree?

A distance of 2 feet from the landing spot is 28 feet from the launch site.

Substitute 28 for x to determine how high the rocket will be when it has traveled 28 feet from the launch site.

$$f(x)=-\frac{16}{9}(x-15)^2+400$$

$$f(28)=-\frac{16}{9}(28-15)^2+400$$

$$f(28)=-\frac{16}{9}(13)^2+400$$

$$f(28)=-\frac{16}{9}(169)+400$$

$$f(28)=99\frac{5}{9}$$

The height of the rocket when it is 2 feet from the landing spot will be approximately 99.6 feet. The rocket will not make it over the tree that stands 115 feet tall.

A–SSE.3c* • Algebra

Writing Exponential Expressions in Equivalent Forms

Common Core State Standard

A–SSE.3

Choose and produce an equivalent form of an expression to reveal and explain properties of the quantity represented by the expression.*

c. Use the properties of exponents to transform expressions for exponential functions. *For example the expression 1.15^t can be rewritten as $(1.15^{1/12})^{12t} \approx 1.012^{12t}$ to reveal the approximate equivalent monthly interest rate if the annual rate is 15%.*

Problem-Based Task: Options of Interest

Devin has money to invest and is considering two investment options as described below.

- Option A: Deposit money in a bank account, which pays 6% annual interest, compounded annually.
- Option B: Lend money to his nephew Hudson, who owns a business and needs capital. Hudson will pay 0.5% monthly interest, compounded monthly.

Use approximate equivalent monthly interest rates, compounded monthly, to determine which option would result in the greatest amount of money earned for a $20,000 investment over 5 years.

Problem-Based Task: Options of Interest

Coaching

a. What is the general exponential growth model?

b. The account in Option A pays 6% annual interest, compounded annually. How can you represent that fact in the exponential growth model, letting t represent a number of years?

c. What is the answer to part b in simplified form?

d. The exponential expression written in part c represents a growth rate per year for t years. How much time is t years, expressed in months?

e. How can you rewrite the exponential expression $(1.06)^t$ so that it represents a growth rate per month for t years expressed in months?

f. How can the Power of a Power Property be used to simplify your answer to part e?

g. What does the exponential expression in part f represent?

h. The given rate for Option B is 0.5% monthly interest, compounded monthly. How can you represent that fact in the exponential growth model, letting t represent a number of months?

i. What is the answer to part h in simplified form?

j. How can you use the general exponential growth model to find the value, rounded to the nearest cent, of $20,000 invested for 5 years at the monthly interest rate, compounded monthly, described in part g?

k. How can you use the general exponential growth model to find the value, rounded to the nearest cent, of $20,000 invested for 5 years at 0.5% monthly interest, compounded monthly?

l. Which option would result in the greatest amount of money earned for a $20,000 investment over 5 years?

Problem-Based Task: Options of Interest

Coaching Sample Responses

a. What is the general exponential growth model?

The general exponential growth model is $f(t) = a(1 + r)^t$.

b. The account in Option A pays 6% annual interest, compounded annually. How can you represent that fact in the exponential growth model, letting *t* represent a number of years?

If the annual interest rate is 6%, then let $r = 0.06$. This rate can be represented as $f(t) = a(1 + 0.06)^t$.

c. What is the answer to part b in simplified form?

$f(t) = a(1.06)^t$

d. The exponential expression written in part c represents a growth rate per year for *t* years. How much time is *t* years, expressed in months?

There are 12 months per year, so 12*t* months represents *t* years expressed in months.

e. How can you rewrite the exponential expression $(1.06)^t$ so that it represents a growth rate per month for *t* years expressed in months?

Rewrite the exponential expression $(1.06)^t$ so that the exponent is 12*t*.

Write the exponent *t* as the product of 12 and its reciprocal, $\frac{1}{12}$.

$$1.06^t = 1.06^{\frac{1}{12} \bullet 12t} = \left(1.06^{\frac{1}{12}}\right)^{12t}$$

f. How can the Power of a Power Property be used to simplify your answer to part e?

Use a calculator to approximate $(1.06)^{\frac{1}{12}}$ and substitute that value into the expression from part e.

$$1.06^t = 1.06^{\frac{1}{12} \bullet 12t} = \left(1.06^{\frac{1}{12}}\right)^{12t} \approx (1.004868)^{12t}$$

g. What does the exponential expression in part f represent?

The exponential expression represents 0.4868% monthly interest, compounded monthly.

h. The given rate for Option B is 0.5% monthly interest, compounded monthly. How can you represent that fact in the exponential growth model, letting t represent a number of months?

If the monthly interest is 0.5%, then let $r = 0.005$. This rate can be represented as $f(t) = a(1 + 0.005)^t$.

i. What is the answer to part h in simplified form?

$f(t) = a(1.005)^t$

j. How can you use the general exponential growth model to find the value, rounded to the nearest cent, of \$20,000 invested for 5 years at the monthly interest rate, compounded monthly, described in part g?

The value of \$20,000 invested for 5 years at the monthly interest rate of 0.4868%, compounded monthly, is found using the following expression:

$f(t) = a(1 + r)^t = 20{,}000(1 + 0.004868)^{60} = 20{,}000(1.004868)^{60} \approx \$26{,}765.23$

The value is approximately \$26,765.23.

k. How can you use the general exponential growth model to find the value, rounded to the nearest cent, of \$20,000 invested for 5 years at 0.5% monthly interest, compounded monthly?

The value of \$20,000 invested for 5 years at 0.5% monthly interest, compounded monthly, is found using the following expression:

$f(t) = a(1 + r)^t = 20{,}000(1 + 0.005)^{60} = 20{,}000(1.005)^{60} \approx \$26{,}977.00$

The value is approximately \$26,977.00.

l. Which option would result in the greatest amount of money earned for a \$20,000 investment over 5 years?

Option B would result in the greatest amount of money earned for a \$20,000 investment over 5 years.

Common Core State Standard

A–APR.1

Understand that polynomials form a system analogous to the integers, namely, they are closed under the operations of addition, subtraction, and multiplication; add, subtract, and multiply polynomials.

Problem-Based Task: Cabin Perimeter

Soren has been hired to design a small cabin. He has drawn the blueprint below. His client is still determining the overall size of the cabin, but Soren has labeled the known lengths in feet. He wants to find an expression to represent the perimeter of the entire space. The perimeter of the cabin is the sum of all four sides and can be written as perimeter = $2a + 2b$. Find an expression in terms of x that shows the total perimeter.

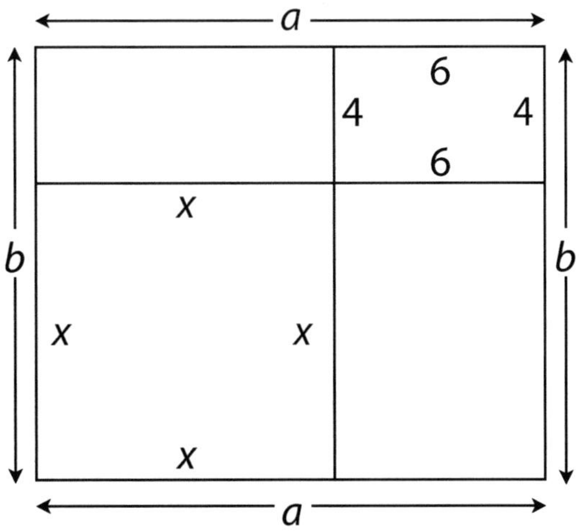

Problem-Based Task: Cabin Perimeter

Coaching

a. What does a equal in terms of x?

b. What does b equal in terms of x?

c. Using the formula for perimeter and expressions for a and b, write a formula for the cabin's perimeter in terms of x.

d. What is the simplified perimeter of the cabin in terms of x?

Problem-Based Task: Cabin Perimeter

Coaching Sample Responses

a. What does a equal in terms of x?

The length of a contains two known lengths: x and 6 feet.

The sum of these two lengths equals a: $(x + 6)$ feet.

b. What does b equal in terms of x?

The length of b contains two known lengths: x and 4 feet.

The sum of these two lengths equals b: $(x + 4)$ feet.

c. Using the formula for perimeter and expressions for a and b, write a formula for the cabin's perimeter in terms of x.

Replace a and b in the formula with the determined expressions.

perimeter $= 2a + 2b$

perimeter $= 2(x + 6) + 2(x + 4)$

d. What is the simplified perimeter of the cabin in terms of x?

perimeter $= 2(x + 6) + 2(x + 4)$

perimeter $= 2x + 12 + 2(x + 4)$

perimeter $= 2x + 12 + 2x + 8$

perimeter $= 2x + 2x + 12 + 8$

perimeter $= 4x + 20$

The simplified perimeter of the cabin in terms of x is $(4x + 20)$ feet.

A–APR.1 • Algebra

Multiplying Polynomials

Common Core State Standard

A–APR.1

Understand that polynomials form a system analogous to the integers, namely, they are closed under the operations of addition, subtraction, and multiplication; add, subtract, and multiply polynomials.

Problem-Based Task: Architectural Area

An architect is creating a template, or reusable pattern, of the design of a bathroom. One part of the bathroom has a standard size in order to fit a standard bathtub, and one part of the bathroom can vary based on what the customer wants. The architect's template is shown below, and all units are in inches. The area of a rectangle is *lw*, or in this case, area = *ab*. Find an expression to determine the total area of the bathroom for any value of *x*.

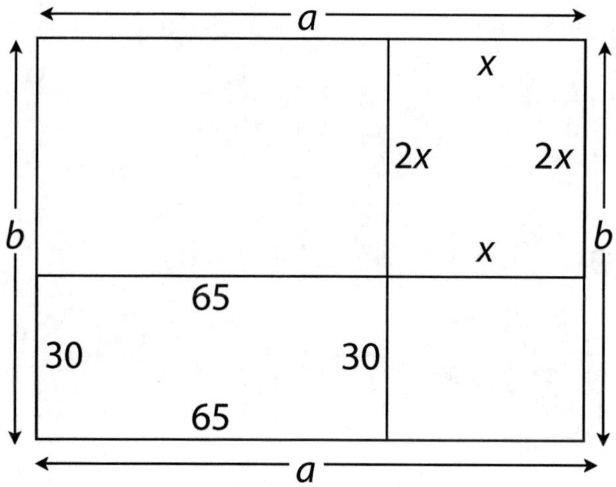

Problem-Based Task: Architectural Area

Coaching

a. What does a equal in terms of x?

b. What does b equal in terms in x?

c. Use the formula for area and expressions for a and b to write a formula for the area of the bathroom.

d. What is the simplified area of the bathroom written in terms of x?

Problem-Based Task: Architectural Area

Coaching Sample Responses

a. What does a equal in terms of x?

The length of a contains two known lengths: x and 65 inches.

The sum of these two lengths equals a: $(x + 65)$ inches.

b. What does b equal in terms in x?

The length of b contains two known lengths: $2x$ and 30 inches.

The sum of these two lengths equals b: $(2x + 30)$ inches.

c. Use the formula for area and expressions for a and b to write a formula for the area of the bathroom.

Replace a and b in the formula with the determined expressions.

$$\text{area} = ab = (x + 65)(2x + 30)$$

d. What is the simplified area of the bathroom written in terms of x?

Find the product of the two polynomials.

Find the product of the first terms, outside terms, inside terms, and last terms.

$$(x + 65)(2x + 30) = 2x^2 + 30x + 130x + 1950$$

Simplify the expression by adding the two like terms, each containing x to the first power.

$$2x^2 + 30x + 130x + 1950 = 2x^2 + 160x + 1950$$

The simplified area of the bathroom written in terms of x is $(2x^2 + 160x + 1950)$ inches^2.

> **Common Core State Standard**
>
> **A–CED.1**
>
> Create equations and inequalities in one variable and use them to solve problems. *Include equations arising from linear and quadratic functions, and simple rational and exponential functions.*★

Problem-Based Task: Golden Rectangles and the Golden Ratio

Golden rectangles have been used by artists, architects, and other creative individuals for hundreds of years because of a unique property they all share. A golden rectangle is a rectangle that can be divided into a square and a smaller rectangle that is similar to the original rectangle. In the diagram below, quadrilateral *ABCD* is a golden rectangle, and $\overline{PQ}$ divides it into square *APQD* and rectangle *BCQP*, which is similar to quadrilateral *ABCD*. (Using the similarity symbol, $BCQP \sim ABCD$).

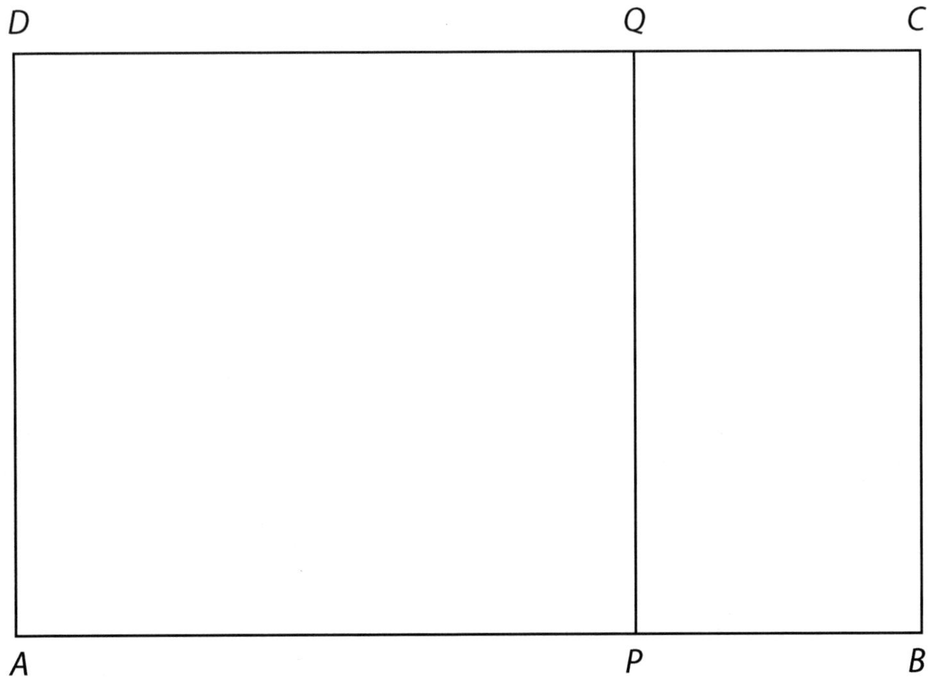

The ratio of length to width in a golden rectangle is called the golden ratio. To find the golden ratio, you can let the width be 1, let the length be *x*, and then find the value of *x* (because $\frac{\text{length}}{\text{width}} = \frac{x}{1} = x$). Let $AD = 1$ and $AB = x$. Find the golden ratio.

Problem-Based Task: Golden Rectangles and the Golden Ratio

Coaching

a. What is true about corresponding sides of similar polygons?

b. You know that $BCQP \sim ABCD$. How can you complete the proportion $\frac{AB}{BC} = \frac{?}{?}$, using $\overline{AD}$ and $\overline{BP}$?

c. You know $AD = 1$ and $AB = x$. What is the length of $\overline{BC}$? Explain.

d. What is the length of $\overline{AP}$? Explain.

e. What is the length of $\overline{BP}$? Explain your answer and then sketch the given diagram, labeling lengths $\overline{AB}$, $\overline{AD}$, $\overline{AP}$, $\overline{BC}$, and $\overline{BP}$.

f. How can you use the proportion from part b to find the value of *x*? Show your work and explain your steps.

g. What is a decimal approximation of the golden ratio?

h. What does the decimal approximation of the golden ratio tell you about the shape of any golden rectangle?

Problem-Based Task: Golden Rectangles and the Golden Ratio

Coaching Sample Responses

a. What is true about corresponding sides of similar polygons?

They are proportional.

b. You know that $BCQP \sim ABCD$. How can you complete the proportion $\frac{AB}{BC}=\frac{?}{?}$, using $\overline{AD}$ and $\overline{BP}$?

$$\frac{AB}{BC}=\frac{AD}{BP}$$

c. You know $AD = 1$ and $AB = x$. What is the length of $\overline{BC}$? Explain.

$BC = AD$ because opposite sides of a rectangle are congruent; therefore, $BC = 1$.

d. What is the length of $\overline{AP}$? Explain.

$AD = 1$ and $APQD$ is a square; therefore, $AP = 1$.

e. What is the length of $\overline{BP}$? Explain your answer and then sketch the given diagram, labeling lengths $\overline{AB}$, $\overline{AD}$, $\overline{AP}$, $\overline{BC}$, and $\overline{BP}$.

$AB = x$, $AP = 1$, and $\overline{BP}$ is the length remaining when 1 is subtracted from x; therefore, $BP = x - 1$.

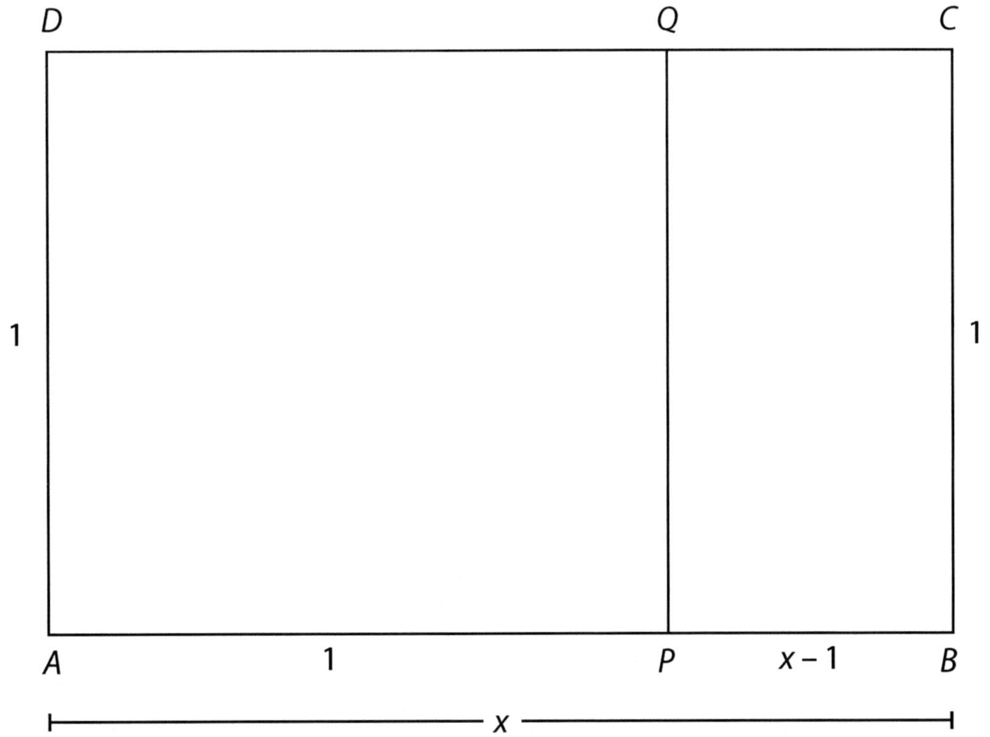

f. How can you use the proportion from part b to find the value of x? Show your work and explain your steps.

$\frac{AB}{BC}=\frac{AD}{BP}$	Write the proportion from part b.
$\frac{(x)}{(1)}=\frac{(1)}{(x-1)}$	Substitute the values and expressions previously identified.
$(x-1)\left(\frac{x}{1}\right)=(x-1)\left(\frac{1}{x-1}\right)$	Multiply both sides by the LCD.
$(x-1)(x)=1$	Simplify.
$x^2-x=1$	Distribute.
$x^2-x-1=0$	Subtract 1 from both sides.
$x=\frac{-(-1)\pm\sqrt{(-1)^2-4(1)(-1)}}{2(1)}$	Use the quadratic formula, $x=\frac{-b\pm\sqrt{b^2-4ac}}{2a}$.
$x=\frac{1\pm\sqrt{1+4}}{2}$	Simplify.
$x=\frac{1+\sqrt{5}}{2}$ or $x=\frac{1-\sqrt{5}}{2}$	

The value of $\frac{1-\sqrt{5}}{2}$ is negative, so it cannot represent the length of a rectangle.

The length of rectangle *ABCD* is $x=\frac{1+\sqrt{5}}{2}$, which is the golden ratio.

g. What is a decimal approximation of the golden ratio?

$\frac{1+\sqrt{5}}{2}$ is approximately equal to 1.618.

h. What does the decimal approximation of the golden ratio tell you about the shape of any golden rectangle?

The length is about 1.618 times the width.

A–CED.1★ • Algebra

Creating Rational Inequalities

Common Core State Standard

A–CED.1

Create equations and inequalities in one variable and use them to solve problems. *Include equations arising from linear and quadratic functions, and simple rational and exponential functions.*★

Problem-Based Task: Capping Costs

Verona sells sporting goods. One product she sells is a baseball cap with a team letter of the customer's choice. Her startup cost for the equipment to attach the letters to the caps is \$200. It costs her \$5 to process each cap, which includes what she pays for the cap. She sells each cap for \$15. Verona's goal is for her average profit per cap to exceed her average cost per cap. How many caps could she sell to achieve her goal? What will be the minimum amount of profit she will earn if she achieves her goal?

Problem-Based Task: Capping Costs

Coaching

a. What are the two quantities you need to compare in order to answer the questions?

b. What is the rule for the function $C(x)$ that represents the cost in dollars to process x caps, including the startup cost?

c. What is the rule for the function $C_A(x)$ that represents the average cost in dollars per cap to process x caps?

d. What is the rule for the function $R(x)$ that represents the revenue in dollars that Verona takes in by selling x caps?

e. What is the rule for the function $P(x)$ that represents the profit in dollars that Verona earns by selling x caps?

f. What is the rule for the function $P_A(x)$ that represents the average profit in dollars per cap that Verona earns by selling x caps?

g. What is Verona's goal?

h. How can you use the answers to parts c and f to write a rational inequality that represents Verona's goal?

i. How can you find the positive solutions of the inequality? Explain your method.

j. How many caps could Verona sell to achieve her goal of having her average profit per cap exceed her average cost per cap?

k. What will be the minimum amount of profit Verona will earn if she achieves her goal? How can you justify your answer?

Problem-Based Task: Capping Costs

Coaching Sample Responses

a. What are the two quantities you need to compare in order to answer the questions?

The two quantities that need to be compared are average profit per cap and average cost per cap.

b. What is the rule for the function $C(x)$ that represents the cost in dollars to process x caps, including the startup cost?

The rule for the function $C(x)$ that represents the cost in dollars to process x caps is $C(x) = 200 + 5x$.

c. What is the rule for the function $C_A(x)$ that represents the average cost in dollars per cap to process x caps?

The rule for the function $C_A(x)$ that represents the average cost in dollars per cap is $C_A(x)=\frac{200+5x}{x}$.

d. What is the rule for the function $R(x)$ that represents the revenue in dollars that Verona takes in by selling x caps?

The rule for the function $R(x)$ that represents the revenue in dollars is $R(x) = 15x$.

e. What is the rule for the function $P(x)$ that represents the profit in dollars that Verona earns by selling x caps?

The rule for the function $P(x)$ that represents the profit in dollars is as follows:

$$P(x) = R(x) - C(x)$$

$$P(x) = 15x - (200 + 5x)$$

$$P(x) = 10x - 200$$

f. What is the rule for the function $P_A(x)$ that represents the average profit in dollars per cap that Verona earns by selling x caps?

The rule for the function $P_A(x)$ that represents the average profit in dollars per cap is $P_A(x)=\frac{10x-200}{x}$.

g. What is Verona's goal?

Verona's goal is to have her average profit per cap exceed her average cost per cap.

h. How can you use the answers to parts c and f to write a rational inequality that represents Verona's goal?

Verona wants $P_A(x) > C_A(x)$, so the inequality is $\frac{10x-200}{x} > \frac{200+5x}{x}$.

i. How can you find the positive solutions of the inequality? Explain your method.

Solve the rational inequality.

$\frac{10x-200}{x} > \frac{200+5x}{x}$	Write the inequality.
$10x - 200 > 200 + 5x$	Multiply both sides by x. The inequality symbol stays the same because x is a number of caps and is therefore positive.
$5x - 200 > 200$	Subtract $5x$ from both sides.
$5x > 400$	Add 200 to both sides.
$x > 80$	Divide both sides by 5.

j. How many caps could Verona sell to achieve her goal of having her average profit per cap exceed her average cost per cap?

Verona could sell 81 or more caps to achieve her goal.

k. What will be the minimum amount of profit Verona will earn if she achieves her goal? How can you justify your answer?

The minimum amount of profit Verona will earn is $610. Verona must sell a minimum of 81 caps, and $P(81) = 10(81) - 200 = 610$.

NAME:

A–CED.2★ • Algebra

Graphing Rational Equations

Common Core State Standard

A–CED.2

Create equations in two or more variables to represent relationships between quantities; graph equations on coordinate axes with labels and scales.★

Problem-Based Task: Lacrosse Stats

Latonya is the goaltender on her school's lacrosse team. She blocked 2 shots out of 5 attempted shots on goal in her first game this season, so her save percentage is now 40%. She is interested in knowing how her save percentage will change if she is able to make saves from now on without missing. A save percentage is a ratio, written in percent form. A rational equation is used to represent Latonya's save ratio function $f(x)$, where x is the number of consecutive successful saves she makes from now on. What is the meaning of the function's end behavior in the context of the problem situation?

Problem-Based Task: Lacrosse Stats

Coaching

a. What is a save ratio, using the terms "saves" and "shots on goal?"

b. What is Latonya's save ratio now? Explain.

c. What will be her save ratio when she makes x consecutive saves?

d. What is the rational function $f(x)$ that represents Latonya's save ratio, where x is the number of consecutive saves she successfully makes from now on?

e. To graph the function, it will help to identify the asymptote(s). What value of a denominator in a rational function rule is associated with a vertical asymptote?

f. What value of x makes the denominator equal to 0?

g. What is the equation of the vertical asymptote?

h. You need to draw the asymptote and then plot some points on both sides of the asymptote. What are the points on the graph that have the x-values –8, –7, –6, –4, –3, and –2?

i. How can you use the information from parts g and h to help sketch the function graph?

j. Sketch the graph, but not very far left of (–8, 2) or very far right of (–2, 0).

k. What are the points on the graph that have x-values –60, –40, –20, 20, 40, and 60? Round to the nearest thousandth if you need to round.

l. What value are the y-values getting close to as the x-values get ever farther away from 0 on the left and right (that is, as the x-values decrease and increase without bound)?

m. The function is $f(x)=\dfrac{x+2}{x+5}$. Is it possible that $f(x) = 1$? Explain.

n. The line $y = b$ is a horizontal asymptote if $f(x)$ gets closer to b as x either increases or decreases without bound. What is the horizontal asymptote for this function?

o. How can you use the information in part n to finish sketching the graph of the function?

p. Finish sketching the graph of the function.

q. What part of the graph is relevant to the situation? Explain.

r. What is the meaning of the function's end behavior in the context of the problem situation?

Problem-Based Task: Lacrosse Stats

Coaching Sample Responses

a. What is a save ratio, using the terms "saves" and "shots on goal?"

The save ratio can be written as $\frac{\text{saves}}{\text{shots on goal}}$.

b. What is Latonya's save ratio now? Explain.

Latonya's save ratio is $\frac{\text{saves}}{\text{shots on goal}} = \frac{2}{5}$. This can be written as 0.4 as well as 40%.

c. What will be her save ratio when she makes x consecutive saves?

Her save ratio can be written as $\frac{2+x}{5+x}$ when she makes x consecutive successful attempts.

d. What is the rational function $f(x)$ that represents Latonya's save ratio, where x is the number of consecutive saves she successfully makes from now on?

The rational equation that represents her save ratio is $f(x)=\frac{2+x}{5+x}$.

e. To graph the function, it will help to identify the asymptote(s). What value of a denominator in a rational function rule is associated with a vertical asymptote?

The value of 0 in the denominator of a rational function is associated with the vertical asymptote.

f. What value of x makes the denominator equal to 0?

Set the denominator equal to 0 and solve for x to find the value that makes the denominator equal to 0.

$5 + x = 0$

$x = -5$

The value of x that makes the denominator equal to 0 is –5.

g. What is the equation of the vertical asymptote?

The equation of the vertical asymptote is $x = -5$.

h. You need to draw the asymptote and then plot some points on both sides of the asymptote. What are the points on the graph that have the x-values –8, –7, –6, –4, –3, and –2?

By solving for $f(x)$, the points are (–8, 2), (–7, 2.5), (–6, 4), (–4, –2), (–3, –0.5), and (–2, 0).

i. How can you use the information from parts g and h to help sketch the function graph?

Draw the asymptote, plot the points, and then sketch the function graph through the plotted points.

j. Sketch the graph, but not very far left of (–8, 2) or very far right of (–2, 0).

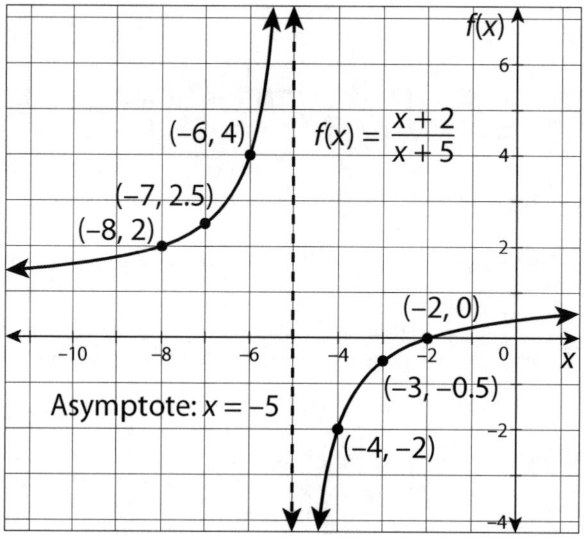

k. What are the points on the graph that have *x*-values –60, –40, –20, 20, 40, and 60? Round to the nearest thousandth if you need to round.

Use the equation for *f*(*x*) to calculate each point. The points are (–60, 1.055), (–40, 1.086), (–20, 1.2), (20, 0.88), (40, 0.933), and (60, 0.954).

l. What value are the *y*-values getting close to as the *x*-values get ever farther away from 0 on the left and right (that is, as the *x*-values decrease and increase without bound)?

The *y*-values are getting closer to 1.

m. The function is $f(x)=\frac{x+2}{x+5}$. Is it possible that *f*(*x*) = 1? Explain.

No, it is not possible. The numerator and denominator are never equal; therefore, there is no value of *x* that makes $x + 2 = x + 5$.

n. The line $y = b$ is a horizontal asymptote if *f*(*x*) gets closer to *b* as *x* either increases or decreases without bound. What is the horizontal asymptote for this function?

The horizontal asymptote for this function is $y = 1$.

o. How can you use the information in part n to finish sketching the graph of the function?

Draw the horizontal asymptote $y = 1$ and extend the curved branches of the function graph to the left and to the right, showing that they get close to the asymptote but do not touch it.

p. Finish sketching the graph of the function.

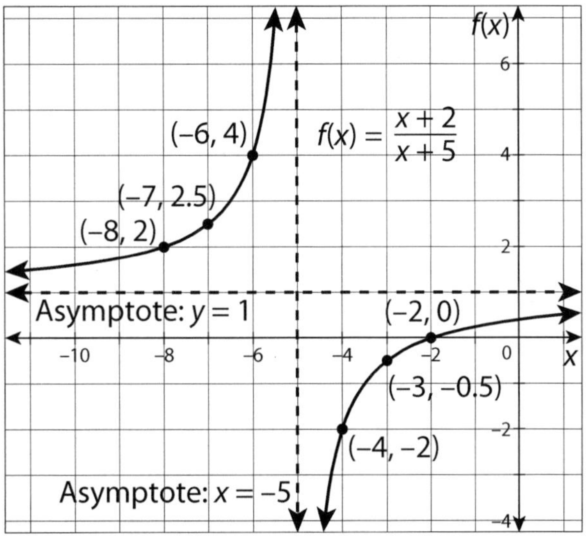

q. What part of the graph is relevant to the situation? Explain.

The relevant part of the graph is the intersection with the y-axis and all points to the right that have integer x-values. Explanation: x represents the number of consecutive successful saves she makes from now on, so x must be a whole number. That is, x must be in the set $\{0, 1, 2, 3, \ldots\}$.

r. What is the meaning of the function's end behavior in the context of the problem situation?

The description of the end behavior is: $f(x)$ gets closer to 1, but is always less than 1, as x increases without bound. In the context of the problem situation, this means that Latonya's save percentage gets closer to 100%, but is always less than 100%, as she makes ever more consecutive successful saves.

A–CED.4★ • Algebra

Rearranging Formulas

Common Core State Standard

A–CED.4

Rearrange formulas to highlight a quantity of interest, using the same reasoning as in solving equations. *For example, rearrange Ohm's law $V = IR$ to highlight resistance R.*★

Problem-Based Task: Selecting a Cylinder Size

A local soda company is investigating several different models for a cylindrical, recyclable can for its organic root beer. The formula for the volume of a cylinder is $V = \pi r^2 h$, where πr^2 is the area of the cylinder's base and h is the height of the cylinder. If the volume of each can needs to be approximately 12.56 in^2, and the height of the can needs to be 4 inches, what should be the radius of each can?

Problem-Based Task: Selecting a Cylinder Size

Coaching

a. What do each of the variables in the formula $V = \pi r^2 h$ represent?

b. What is the focus of the given formula? Explain.

c. What should the new focus be? Explain.

d. Solve the formula for the unknown variable.

e. Is the negative disregarded in this situation? Explain.

f. What values are given in the problem?

g. What should be the radius of each can if the volume needs to be approximately 12.56 in^2 and the height needs to be 4 inches?

Problem-Based Task: Selecting a Cylinder Size

Coaching Sample Responses

a. What do each of the variables in the formula $V = \pi r^2 h$ represent?

V is the volume of the can, *r* is the radius of the can's bases, and *h* is the height of the can.

b. What is the focus of the given formula? Explain.

The focus of the given formula is the volume because *V* is isolated.

c. What should the new focus be? Explain.

The new focus should be the radius, *r*, because this is the value the company is trying to determine.

d. Solve the formula for the unknown variable.

Solve the formula for the radius, *r*.

$$\frac{V}{\pi h} = \frac{\pi r^2 h}{\pi h}$$

$$\sqrt{\frac{V}{\pi h}} = \sqrt{r^2}$$

$$r = \pm\sqrt{\frac{V}{\pi h}}$$

e. Is the negative disregarded in this situation? Explain.

Yes. The radius is a distance, so the equation is $r = \sqrt{\frac{V}{\pi h}}$.

f. What values are given in the problem?

The volume is 12.56 in^2 and the height is 4 inches.

g. What should be the radius of each can if the volume needs to be approximately 12.56 in² and the height needs to be 4 inches?

Substitute 4 for the height and solve the equation for *r*.

$$r=\sqrt{\frac{12.56}{\pi(4)}}$$

$$r=\sqrt{\frac{12.56}{4\pi}}$$

$$r=\sqrt{1}$$

$$r=1$$

The radius of each can should be 1 inch.

A–CED.1★; A–REI.4b • Algebra

Taking the Square Root of Both Sides

Common Core State Standards

A–CED.1

Create equations and inequalities in one variable and use them to solve problems. *Include equations arising from linear and quadratic functions, and simple rational and exponential functions.*★

A–REI.4

Solve quadratic equations in one variable.

b. Solve quadratic equations by inspection (e.g., for $x^2 = 49$), taking square roots, completing the square, the quadratic formula and factoring, as appropriate to the initial form of the equation. Recognize when the quadratic formula gives complex solutions and write them as $a \pm bi$ for real numbers a and b.

Problem-Based Task: Time to Splash

Nina dives into a pool from a platform 3.75 feet above the water. Her height above the water in feet x seconds into the jump is given by the expression $-5(x - 0.5)^2 + 5$. How long will it take Nina to hit the water?

Problem-Based Task: Time to Splash

Coaching

a. What will Nina's height be when she hits the water?

b. Write an equation using your answer from part a.

c. How can you solve your equation from part b?

d. Does a negative answer make sense for this problem? Explain your reasoning.

e. Is your positive answer reasonable for this problem? Explain your reasoning.

f. How long will it take Nina to hit the water?

Problem-Based Task: Time to Splash

Coaching Sample Responses

a. What will Nina's height be when she hits the water?

When Nina hits the water, she will be 0 feet above the water.

b. Write an equation using your answer from part a.

Set the expression for height equal to 0.

$-5(x - 0.5)^2 + 5 = 0$

c. How can you solve your equation from part b?

Solve the equation by isolating the squared binomial and taking the square root of both sides.

First, subtract 5 from both sides.

$-5(x - 0.5)^2 = -5$

Next, divide both sides by –5.

$(x - 0.5)^2 = 1$

Take the square root of both sides.

$x - 0.5 = \pm 1$

$x = 0.5 \pm 1$

$x = 1.5$ and $x = -0.5$

d. Does a negative answer make sense for this problem? Explain your reasoning.

The x stands for the number of seconds after Nina jumps. A negative answer does not make sense for this problem since time cannot be negative, and Nina will not hit the water before she jumps from the platform.

e. Is your positive answer reasonable for this problem? Explain your reasoning.

The positive answer is 1.5 seconds. It is reasonable that Nina would hit the surface of the water after 1.5 seconds.

f. How long will it take Nina to hit the water?

Nina will hit the water after 1.5 seconds.

A–SSE.2; A–CED.1★; A–REI.4b • Algebra

Factoring

Common Core State Standards

A–SSE.2

Use the structure of an expression to identify ways to rewrite it. *For example, see $x^4 - y^4$ as $(x^2)^2 - (y^2)^2$, thus recognizing it as a difference of squares that can be factored as $(x^2 - y^2)(x^2 + y^2)$.*

A–CED.1

Create equations and inequalities in one variable and use them to solve problems. *Include equations arising from linear and quadratic functions, and simple rational and exponential functions.*★

A–REI.4

Solve quadratic equations in one variable.

b. Solve quadratic equations by inspection (e.g., for $x^2 = 49$), taking square roots, completing the square, the quadratic formula and factoring, as appropriate to the initial form of the equation. Recognize when the quadratic formula gives complex solutions and write them as $a \pm bi$ for real numbers a and b.

Problem-Based Task: On the Shelf

Kareem wants to put up a small corner shelf in his room. The shelf is in the shape of a right triangle. One leg of the triangle is 2 inches shorter than the other. The longest side of the shelf is 10 inches long. Kareem needs to find the area of the shelf so that he knows if he has enough paint to cover the top of the shelf. The area of a triangle is $A = \frac{1}{2}bh$.

Problem-Based Task: On the Shelf

Coaching

a. Let the length of the long leg of the triangle equal x. How can the length of the short leg be expressed in terms of x?

b. How can the Pythagorean Theorem help us to relate the lengths of all three sides of the triangular shelf?

c. How can your equation be simplified and set equal to 0?

d. How can the quadratic expression be factored?

e. How can the Zero Product Property help us to solve the factored equation?

f. Do both of your answers make sense as side lengths?

g. What are the lengths of the two legs of the triangle?

h. What is the area of the shelf?

Problem-Based Task: On the Shelf

Coaching Sample Responses

a. Let the length of the long leg equal x. How can the length of the short leg be expressed in terms of x?

The short leg is $(x - 2)$ inches long.

b. How can the Pythagorean Theorem help us to relate the lengths of all three sides of the triangular shelf?

Substitute the expressions into the Pythagorean Theorem to determine the actual lengths.

$x^2 + (x - 2)^2 = 10^2$

c. How can your equation be simplified and set equal to 0?

Square the binomial term, combine like terms, and then subtract 100 from both sides.

$x^2 + x^2 - 4x + 4 = 100$

$2x^2 - 4x + 4 = 100$

$2x^2 - 4x - 96 = 0$

d. How can the quadratic expression be factored?

Factor out the greatest common factor of 2.

$2(x^2 - 2x - 48) = 0$

Find two numbers whose product is –48 and whose sum is –2.

The numbers are –8 and 6.

Use these numbers to find the factors.

The factors are $(x - 8)$ and $(x + 6)$.

Write the expression as a product of factors.

$2(x - 8)(x + 6) = 0$

e. How can the Zero Product Property help us to solve the factored equation?

Set each factor equal to 0 and solve for x.

$x-8=0$ $\qquad$ $x+6=0$

$x=8$ $\qquad$ $x=-6$

f. Do both of your answers make sense as side lengths?

Side lengths must be positive, so –6 does not make sense.

g. What are the lengths of the two legs of the triangle?

One leg of the triangle is 8 inches.

The other leg is 2 inches shorter, so it is 6 inches long.

h. What is the area of the shelf?

Substitute the known values into the formula for the area of a triangle, $A=\frac{1}{2}bh$.

$$A=\frac{1}{2}(8)(6)=24$$

The area of the shelf is 24 square inches.

A–SSE.2; A–CED.1★; A–REI.4a; A–REI.4b • Algebra

Completing the Square

Common Core State Standards

A–SSE.2

Use the structure of an expression to identify ways to rewrite it. *For example, see $x^4 - y^4$ as $(x^2)^2 - (y^2)^2$, thus recognizing it as a difference of squares that can be factored as $(x^2 - y^2)(x^2 + y^2)$.*

A–CED.1

Create equations and inequalities in one variable and use them to solve problems. *Include equations arising from linear and quadratic functions, and simple rational and exponential functions.*★

A–REI.4

Solve quadratic equations in one variable.

a. Use the method of completing the square to transform any quadratic equation in x into an equation of the form $(x - p)^2 = q$ that has the same solutions. Derive the quadratic formula from this form.

b. Solve quadratic equations by inspection (e.g., for $x^2 = 49$), taking square roots, completing the square, the quadratic formula and factoring, as appropriate to the initial form of the equation. Recognize when the quadratic formula gives complex solutions and write them as $a \pm bi$ for real numbers a and b.

Problem-Based Task: Curve Ball

The height of a baseball in feet x seconds after it is thrown is given by $-16x^2 + 32x + 5$. When will the ball be at a height of 7 feet?

Problem-Based Task: Curve Ball

Coaching

a. How can you represent the situation with an equation?

b. Can you combine the two constants?

c. How can you solve the equation?

d. Do both of your answers make sense?

e. When will the ball be at a height of 7 feet?

Problem-Based Task: Curve Ball

Coaching Sample Responses

a. How can you represent the situation with an equation?

Set the expression equal to the given height.

$-16x^2 + 32x + 5 = 7$

b. Can you combine the two constants?

Yes, the two constants can be combined by subtracting 5 from both sides of the equation.

$-16x^2 + 32x = 2$

c. How can you solve the equation?

The equation can be solved by completing the square.

Divide all terms by –16.

$$x^2 - 2x = -\frac{1}{8}$$

The square of half the coefficient of x is 1.

Add 1 to both sides.

$$x^2 - 2x + 1 = \frac{7}{8}$$

Write the left side as the square of a binomial, then solve by taking the square roots.

$$(x-1)^2 = \frac{7}{8}$$

$$x - 1 = \pm\sqrt{\frac{7}{8}}$$

$$x = 1 \pm \sqrt{\frac{7}{8}}$$

$x \approx 1.94$ or $x \approx 0.06$

d. Do both of your answers make sense?

Yes. Both answers make sense. The ball will reach 7 feet on its way up and again on its way down.

e. When will the ball be at a height of 7 feet?

The ball will be at a height of 7 feet after about 0.06 second and again after about 1.94 seconds.

A–CED.1★; A–REI.4a; A–REI.4b • Algebra

Applying the Quadratic Formula

Common Core State Standards

A–CED.1

Create equations and inequalities in one variable and use them to solve problems. *Include equations arising from linear and quadratic functions, and simple rational and exponential functions.*★

A–REI.4

Solve quadratic equations in one variable.

a. Use the method of completing the square to transform any quadratic equation in x into an equation of the form $(x-p)^2 = q$ that has the same solutions. Derive the quadratic formula from this form.

b. Solve quadratic equations by inspection (e.g., for $x^2 = 49$), taking square roots, completing the square, the quadratic formula and factoring, as appropriate to the initial form of the equation. Recognize when the quadratic formula gives complex solutions and write them as $a \pm bi$ for real numbers a and b.

Problem-Based Task: Divine Proportion

The golden ratio is a special number. It is often represented by the Greek letter *phi*, ϕ. Artists and architects have long used the golden ratio to create work that is pleasing to the human eye. The ancient Greeks called the golden ratio the divine proportion; they believed the number to have mystical properties. One property of the golden ratio is that it is the only positive number that when increased by 1 is equal to its square. Find the golden ratio.

Problem-Based Task: Divine Proportion

Coaching

a. What equation can you use to solve for the golden ratio?

b. What form does your equation need to be in so that the quadratic formula can be used? Write your equation in that form.

c. How can you solve your equation using the quadratic formula?

d. Do both of your answers satisfy the given property?

e. What is the golden ratio?

Problem-Based Task: Divine Proportion

Coaching Sample Responses

a. What equation can you use to solve for the golden ratio?

$\phi^2 = \phi + 1$

b. What form does your equation need to be in so that the quadratic formula can be used? Write your equation in that form.

The equation needs to be in standard form.

Move all terms to one side so that the equation can be set equal to 0.

Subtract ϕ and 1 from both sides.

$\phi^2 - \phi - 1 = 0$

c. How can you solve your equation using the quadratic formula?

Find the values of a, b, and c.

$a = 1$, $b = -1$, and $c = -1$

Substitute the values you found into the quadratic formula.

$$\phi = \frac{-(-1) \pm \sqrt{(-1)^2 - 4(1)(-1)}}{2(1)}$$

Simplify.

$$\phi = \frac{1 \pm \sqrt{1+4}}{2}$$

$$\phi = \frac{1 \pm \sqrt{5}}{2} \approx 1.618 \text{ and } \approx -0.618$$

d. Do both of your answers satisfy the given property?

No. The golden ratio must be positive, so the negative solution does not work.

e. What is the golden ratio?

The golden ratio is $\frac{1+\sqrt{5}}{2}$ or approximately 1.618.

Solving Quadratic Inequalities

Common Core State Standards

A–SSE.2

Use the structure of an expression to identify ways to rewrite it. *For example, see* $x^4 - y^4$ *as* $(x^2)^2 - (y^2)^2$, *thus recognizing it as a difference of squares that can be factored as* $(x^2 - y^2)(x^2 + y^2)$.

A–CED.1

Create equations and inequalities in one variable and use them to solve problems. *Include equations arising from linear and quadratic functions, and simple rational and exponential functions.*★

A–REI.4

Solve quadratic equations in one variable.

b. Solve quadratic equations by inspection (e.g., for $x^2 = 49$), taking square roots, completing the square, the quadratic formula and factoring, as appropriate to the initial form of the equation. Recognize when the quadratic formula gives complex solutions and write them as $a \pm bi$ for real numbers a and b.

Problem-Based Task: Dancing for Charity

A school is planning to host a dance with all profits going to charity. The amount of profit is found by subtracting the total costs from the total income. The income from ticket sales can be expressed as $200x - 10x^2$, where x is the cost of a ticket. The costs of putting on the dance can be expressed as $500 + 20x$. What are all the ticket prices that will result in a profit of \$200 or more?

Problem-Based Task: Dancing for Charity

Coaching

a. How can profit be expressed in terms of income, I, and costs, C?

b. How can profit be expressed in terms of the cost of a ticket, x?

c. How can your expression for profit in terms of ticket cost be used to write an inequality that represents the situation?

d. How can you solve your inequality for x?

e. Do your answers make sense?

f. What are all the ticket prices that will result in a profit of $200 or more?

Problem-Based Task: Dancing for Charity

Coaching Sample Responses

a. How can profit *P* be expressed in terms of income, *I*, and costs, *C*?

Profit is found by subtracting the total costs from the total income.

$P = I - C$

b. How can profit be expressed in terms of the cost of a ticket, *x*?

Substitute the expressions given for income and cost.

$P = (200x - 10x^2) - (500 + 20x)$

Simplify the expression.

$P = 200x - 10x^2 - 500 - 20x$

$P = -10x^2 + 180x - 500$

c. How can your expression for profit in terms of ticket cost be used to write an inequality that represents the situation?

The expression for profit must be greater than or equal to 200.

$-10x^2 + 180x - 500 \geq 200$

d. How can you solve your inequality for *x*?

To set the inequality equal to 0, first subtract 200 from both sides. Change the inequality symbol to an equal sign. Then, divide both sides by 10 to simplify.

$-10x^2 + 180x - 700 \geq 0$

$-10x^2 + 180x - 700 = 0$

$-x^2 + 18x - 70 = 0$

Use the quadratic formula to solve for *x*.

$a = -1$, $b = 18$, and $c = -70$

$$x = \frac{-(18) \pm \sqrt{(18)^2 - 4(-1)(-70)}}{2(-1)}$$

$$x = \frac{-18 \pm \sqrt{324 - 280}}{-2}$$

$$x=\frac{-18\pm\sqrt{44}}{-2}$$

$x\approx 5.68$ or $x\approx 12.32$

Use a number line and test points to solve the inequality.

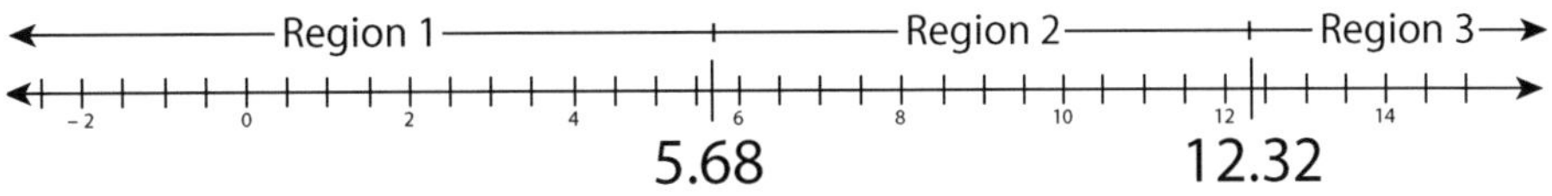

Test $x = 0$ from Region 1:

$-10(0)^2 + 180(0) - 700 = -700$

$-700 \geq 0$ is not true. Therefore, points in Region 1 ($x < 5.68$) are not solutions.

Test $x = 10$ from Region 2:

$-10(10)^2 + 180(10) - 700 = 100$

$100 \geq 0$ is true. Therefore, points in Region 2 ($5.68 \leq x \leq 12.32$) are solutions.

Test $x = 20$ from Region 3:

$-10(20)^2 + 180(20) - 700 = -1100$

$-1100 \geq 0$ is not true. Therefore, points in Region 3 ($x > 12.32$) are not solutions.

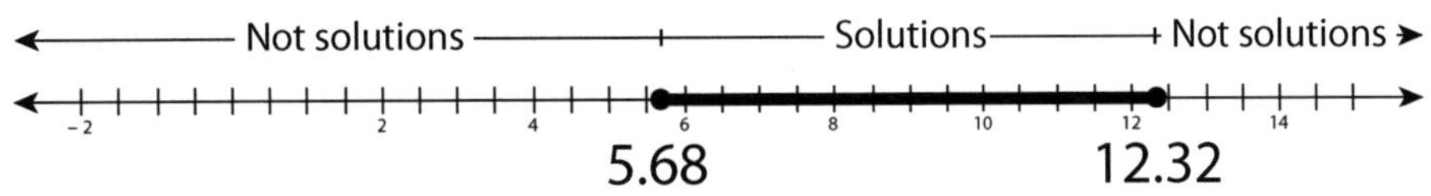

$5.68 \leq x \leq 12.32$

e. Do your answers make sense?

Yes. Ticket prices between \$5.68 and \$12.32 seem reasonable.

f. What are all the ticket prices that will result in a profit of at least \$200?

Profit will be at least \$200 for all ticket prices from \$5.68 to \$12.32, inclusive.

NAME:

A–REI.7 • Algebra

Solving Systems Graphically

Common Core State Standard

A–REI.7

Solve a simple system consisting of a linear equation and a quadratic equation in two variables algebraically and graphically. *For example, find the points of intersection between the line* $y = -3x$ *and the circle* $x^2 + y^2 = 3$.

Problem-Based Task: Constant Acceleration and Constant Speed

A car begins at rest and accelerates. Its distance from its starting position in meters, $C(t)$, can be determined as a function of time in seconds, t, by the formula $C(t) = 2t^2$. A second car, 200 meters ahead, is traveling at a constant speed of 20 meters per second. Its distance from the first car's starting point, $C_2(t)$, in meters can be determined as a function of time, t, in seconds by the formula $C_2(t) = 20t + 200$. How long after the first car accelerates will the cars be side by side?

Problem-Based Task: Constant Acceleration and Constant Speed

Coaching

a. What are the unknowns?

b. What do you need to know in order to answer the question posed in the task?

c. How will you determine the solution?

d. How long after the first car accelerates will the cars be side by side?

Problem-Based Task: Constant Acceleration and Constant Speed

Coaching Sample Responses

a. What are the unknowns?

Time and distance are the unknowns.

b. What do you need to know in order to answer the question posed in the task?

You need to know the time, t, after the first car accelerates until both cars are the same distance from the first car's starting point, $C(t)$.

c. How will you determine the solution?

Graph the functions and find the point(s) of intersection.

d. How long after the first car accelerates will the cars be side by side?

The following graph shows the points of intersection.

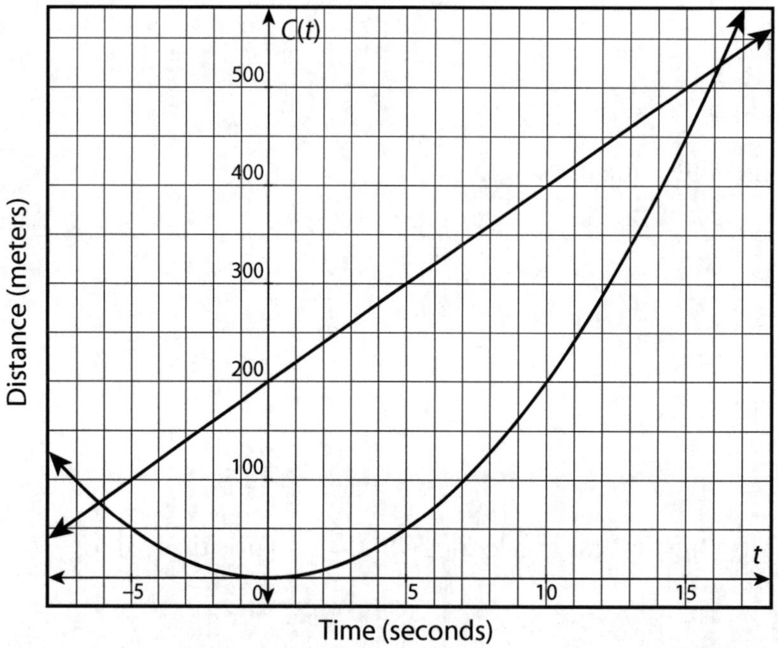

There are two points of intersection, but because the quantities are distance and time, disregard the negative values.

According to the graph, the cars will be side by side about 16.2 seconds after the first car accelerates.

A–REI.7 • Algebra

Solving Systems Algebraically

Common Core State Standard

A–REI.7

Solve a simple system consisting of a linear equation and a quadratic equation in two variables algebraically and graphically. *For example, find the points of intersection between the line* $y = -3x$ *and the circle* $x^2 + y^2 = 3$.

Problem-Based Task: The Business of Bicycles

Nico and Fleur are each designing a fictional bike company for their Business Science class. Nico determines, through research, that the marginal cost of producing a bike is $C_1(x) = 0.0005x^2 + 0.07x + 50$, for which $C_1(x)$ is the marginal cost in dollars per x bikes produced. Fleur determines that the marginal cost for manufacturing her bikes could be modeled by $C_2(x) = \frac{2}{5}x + 62.5$. Assuming each student's fictional company starts "manufacturing" bikes at the same time, at what marginal cost(s) will each company have manufactured the same number of bikes?

Problem-Based Task: The Business of Bicycles

Coaching

a. What are the unknown values?

b. What can you do to solve for the unknowns?

c. How will you solve the quadratic?

d. Is all of the information useful given the context of the problem?

Problem-Based Task: The Business of Bicycles

Coaching Sample Responses

a. What are the unknown values?

The unknowns are x, the number of bikes produced, and $C_1(x)$, the marginal cost of production.

b. What can you do to solve for the unknowns?

Use substitution and set the equations equal to each other.

$$\frac{2}{5}x+62.5=0.0005x^2+0.07x+50$$

c. How will you solve the quadratic?

Rearrange the equation so that one side equals 0.

Since this equation contains both fractions and decimals, it is best to rewrite the equation using just one or the other. Start by rewriting the fraction as a decimal.

$$\frac{2}{5}x+62.5=0.0005x^2+0.07x+50$$

$$0.40x+62.5=0.0005x^2+0.07x+50$$

$$0=0.0005x^2-0.33x-12.5$$

Next, solve the quadratic.

$$x=\frac{-(-0.33)\pm\sqrt{(-0.33)^2-4(0.0005)(-12.5)}}{2(0.0005)}$$

$$x=\frac{0.33\pm\sqrt{0.1339}}{0.001}$$

$x = 695.9235$ and $x = -35.9235$

Substitute in the values of x to find the corresponding y-values.

$$y=\frac{2}{5}(695.9235)+62.5$$

$y = 340.8694$

$$y = \frac{2}{5}(-35.9235) + 62.5$$
$$y = 48.1306$$

The solutions are (695.9235, 340.8694) and (–35.9235, 48.1306).

d. Is all of the information useful given the context of the problem?

Based on the solutions, when the students have spent \$340.87 they will have manufactured about 696 bikes. Also, the students will have manufactured –36 bikes when they have both spent about \$48 on manufacturing. Since it is impossible to manufacture a negative quantity of bikes, the only useful information is the ordered pair (695.9235, 340.8694), which rounds to (696, 340.87).

F–IF.4★ • Functions

Interpreting Key Features of Quadratic Functions

Common Core State Standard

F–IF.4

For a function that models a relationship between two quantities, interpret key features of graphs and tables in terms of the quantities, and sketch graphs showing key features given a verbal description of the relationship. *Key features include: intercepts; intervals where the function is increasing, decreasing, positive, or negative; relative maximums and minimums; symmetries; end behavior; and periodicity.*★

Problem-Based Task: One-on-One Basketball

You and a friend are playing one-on-one basketball at the park. You aim at the hoop and release the ball, which follows a parabolic path. The table below represents the ball's horizontal distance from you versus the height of the ball as it travels toward the center of the hoop, represented by the point (14, 10). Use a quadratic model to determine for what horizontal distances the height of the ball is increasing and decreasing. What is the maximum height that the ball reaches as it continues toward the hoop?

Distance from shooter (feet)	Ball height (feet)
4	10
6	12
12	12
14	10

Problem-Based Task: One-on-One Basketball

Coaching

a. Is there enough information in the table to estimate where the ball is increasing and decreasing, and at what distance it reaches its maximum height?

b. What's the best method to develop a quadratic model?

c. How do you determine when the function is increasing and decreasing?

d. How do you determine the maximum height?

e. What is the maximum height that the ball reaches as it continues toward the hoop?

Problem-Based Task: One-on-One Basketball

Coaching Sample Responses

a. Is there enough information in the table to estimate where the ball is increasing and decreasing, and at what distance it reaches its maximum height?

The heights, or output values, are the symmetric coordinates about some vertical line, the axis of symmetry.

It is difficult to estimate where the function is increasing and decreasing since this depends on the vertex.

b. What's the best method to develop a quadratic model?

A graphing calculator is the best method because it will fill in all of the blanks not included in the table.

The table only gives a snapshot of the basketball's travels along the way.

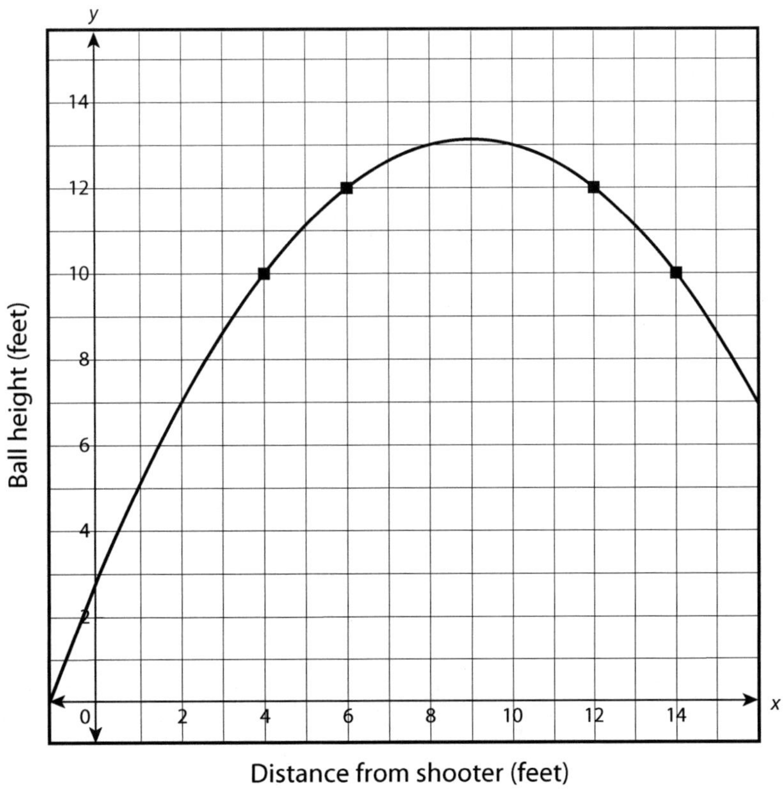

c. How do you determine when the function is increasing and decreasing?

This function has a maximum and therefore is increasing first, then decreasing.

The function will increase until the point of inflection, which appears to happen at a distance of 9 feet.

This can be checked by using the fact that the axis of symmetry should be the middle of the symmetric points.

Choosing the symmetric coordinates of $x = 6$ and $x = 12$, the middle is $x = 9$ because

$$x = \frac{6+12}{2} = \frac{18}{2} = 9.$$

The basketball's height is increasing when its distance is less than 9 feet and decreasing when its distance is greater than 9 feet.

d. How do you determine the maximum height?

The maximum height occurs at the vertex.

Since we've already determined that the basketball's axis of symmetry is at $x = 9$, 9 feet is also the distance that maximizes the height of the ball.

From the model, one can estimate the height to be about 13.1 feet.

However, substituting the x-value into the quadratic model will yield the best estimate.

$$y = -0.125(9)^2 + 2.25(9) + 3 = 13.125$$

e. What is the maximum height that the ball reaches as it continues toward the hoop?

The maximum height the ball reaches is about 13.1 feet.

F–IF.5★ • Functions

Identifying the Domain of a Quadratic Function

Common Core State Standard

F–IF.5

Relate the domain of a function to its graph and, where applicable, to the quantitative relationship it describes. *For example, if the function h(n) gives the number of person-hours it takes to assemble n engines in a factory, then the positive integers would be an appropriate domain for the function.*★

Problem-Based Task: Window Washers

A window washer tosses his wet sponge from a height of 10 meters above ground level to his coworker above him. The sponge reaches its maximum height of 11.25 meters exactly 0.5 second later, but the coworker does not catch the sponge and it falls to ground. After how many seconds will the sponge fall to a height of 10 meters? What is the span of time that the sponge is in the air?

Problem-Based Task: Window Washers

Coaching

a. What are the known values?

b. What are the unknown values?

c. What type of model is needed in order to find the unknowns?

d. What is the vertex form of a quadratic equation?

e. How can the described situation be modeled using the vertex form of a quadratic function?

f. How can you find the value of a?

g. After how many seconds will the sponge fall to a height of 10 meters?

h. What is the reasonable domain of the function?

Problem-Based Task: Window Washers

Coaching Sample Responses

a. What are the known values?

We know the maximum height of the sponge, 11.25 meters.

This value is the vertex and can be written as the point (0.5, 11.25).

We also know the starting point of the sponge, (0, 10).

b. What are the unknown values?

Aside from the sponge's maximum height, we do not know any other heights of the sponge, or the remaining y-values.

We also do not know how long the sponge is in the air at each height.

c. What type of model is needed in order to find the unknowns?

A quadratic model will help determine the unknown values.

d. What is the vertex form of a quadratic equation?

The vertex form of a quadratic function is $y = a(x - h)^2 + k$, where (h, k) is the vertex.

e. How can the described situation be modeled using the vertex form of a quadratic function?

Substitute the vertex and the starting point into the vertex form of a quadratic to get the model.

$y = a(x - h)^2 + k$

$10 = a(0 - 0.5)^2 + 11.25$

f. How can you find the value of a?

Solve the equation for a.

$10 = 0.25a + 11.25$

$-1.25 = 0.25a$

$a = -5$

The quadratic model is $y = -5(x - 0.5)^2 + 11.25$.

g. After how many seconds will the sponge fall to a height of 10 meters?

Use symmetry to determine how many seconds it takes after the sponge is thrown for it to fall to a height of 10 meters.

The point (0, 10) is 0.5 unit to the left of the axis of symmetry; therefore, the point that is 0.5 unit to the right of the axis of symmetry is (1, 10).

The sponge drops to a height of 10 meters 1 second after being thrown.

h. What is the reasonable domain of the function?

The reasonable domain is the interval of time that the sponge is in the air. That can be found from $t = 0$ to the largest x-intercept, where the sponge hits the ground. Using graphing technology, the sponge hits the ground at $t = 2$, or 2 seconds after being tossed. The span of time can be represented as an inequality of $0 < t < 2$.

NAME:

F–IF.6★ • Functions

Identifying the Average Rate of Change

Common Core State Standard

F–IF.6

Calculate and interpret the average rate of change of a function (presented symbolically or as a table) over a specified interval. Estimate the rate of change from a graph.★

Problem-Based Task: Is the Maximum High Enough?

It is Super Bowl season and teams that have made the play-offs have specialists evaluating every aspect of their field game. One particular team received news that their recently injured kicker's field goal kick is modeled by the function $h(t) = -16(x - 1)^2 + 16$, where $h(t)$ is the height of the ball in feet t seconds after it is kicked. If the football needs to clear a 17-foot goalpost, will the ball make it over if this particular team member kicks it? What is the average rate of change of the football's height from the moment it reaches its maximum height to the moment it hits the ground?

Problem-Based Task: Is the Maximum High Enough?

Coaching

a. What are the unknown values?

b. What do you need to know in order to determine whether the football will clear the goalpost?

c. What do you need to know in order to find the average rate of change of the football's height from its maximum height to the ground?

d. If the football needs to clear a 17-foot goalpost, will the ball make it over if this particular team member kicks it?

e. What is the average rate of change of the football's height from the moment it reaches its maximum height to the moment it hits the ground?

Problem-Based Task: Is the Maximum High Enough?

Coaching Sample Responses

a. What are the unknown values?

The maximum height of the football and the average rate of change from the maximum height to the ground are unknown.

b. What do you need to know in order to determine whether the football will clear the goal?

You need to know the y-coordinate of the vertex to determine the maximum height the football reaches.

If the value is greater than 17 feet, then the ball will clear the goalpost.

c. What do you need to know in order to find the average rate of change of the football's height from its maximum height to the ground?

You need to know the x-coordinate of the vertex, $x = 1$, and the other x-intercept to show the moment the ball hits the ground. Use symmetry to determine that the ball will hit the ground when $x = 2$.

d. If the football needs to clear a 17-foot goalpost, will the ball make it over if this particular team member kicks it?

The function is written in vertex form, so the vertex is (1, 16). The maximum height that this player can kick the ball is 16 feet, so the ball will not clear the goalpost.

e. What is the average rate of change of the football's height from the moment it reaches its maximum height to the moment it hits the ground?

The ball hits the ground when $x = 2$, or 2 seconds after big kicked.

The maximum height of the ball occurs at the vertex, when $x = 1$, or 1 second after being kicked.

$h(2) = -16[(2) - 1]^2 + 16 = 0$

$h(1) = -16[(1) - 1]^2 + 16 = 16$

$$\text{Average rate of change} = \frac{0-16}{2-1} = \frac{-16}{1} = -16$$

The average rate of change of the football from its maximum height to the moment it hits the ground is –16 ft/s.

F–IF.7a★ • Functions

Graphing Quadratic Functions

Common Core State Standard

F–IF.7

Graph functions expressed symbolically and show key features of the graph, by hand in simple cases and using technology for more complicated cases.★

a. Graph linear and quadratic functions and show intercepts, maxima, and minima.

Problem-Based Task: How High Can a Frog Jump?

A frog is about to hop from the bank of a creek. The path of the jump can be modeled by the equation $h(x) = -x^2 + 4x + 1$, where $h(x)$ is the frog's height above the water and x is the number of seconds since the frog jumped. A fly is cruising at a height of 5 feet above the water. Is it possible for the frog to catch the fly, given the equation of the frog's jump?

Problem-Based Task: How High Can a Frog Jump?

Coaching

a. Can the frog catch the fly without jumping? How do you know?

b. Is the extremum a maximum or a minimum? Why?

c. What is the vertex of the equation?

d. If the frog jumps to catch the fly, is it possible for the frog to catch the fly?

e. Sketch the graph of the paths of the frog and fly.

Problem-Based Task: How High Can a Frog Jump?

Coaching Sample Responses

a. Can the frog catch the fly without jumping? How do you know?

No, the frog cannot catch the fly without jumping. The frog's initial height is when $t = 0$, represented by the y-intercept.

$h(x) = -x^2 + 4x + 1$

$h(0) = -(0)^2 + 4(0) + 1$

$h(0) = 1$

The frog's initial height above the water is 1 foot when it's sitting on the bank of the creek, so the frog cannot catch a fly that is 5 feet above the water.

b. Is the extremum a maximum or a minimum? Why?

The extremum is a maximum. The value of $a = -1$ and $-1 < 0$, so the parabola opens down.

c. What is the vertex of the equation?

The vertex of the equation can be found by completing the square or by using $\left(\frac{-b}{2a}, f\left(\frac{-b}{2a}\right)\right)$.

Use the original equation $h(x) = -x^2 + 4x + 1$, where $a = -1$ and $b = 4$, to find the x-value of the vertex.

$x = \frac{-b}{2a}$

$x = \frac{-(4)}{2(-1)}$

$x = 2$

The x-coordinate of the vertex is 2.

To find the y-coordinate of the vertex, substitute 2 into the original equation.

$h(x) = -(2)^2 + 4(2) + 1$

$h(x) = 5$

The vertex of the equation is (2, 5).

d. If the frog jumps to catch the fly, is it possible for the frog to catch the fly?

Yes. The maximum height of the frog is the y-value of the vertex, 5. It is possible for the frog to catch the fly because the frog can jump as high or higher than the fly.

e. Sketch the graph of the paths of the frog and fly.

Plot the points of the parabola and connect them with a smooth curve. The fly's path is a horizontal line at $y = 5$.

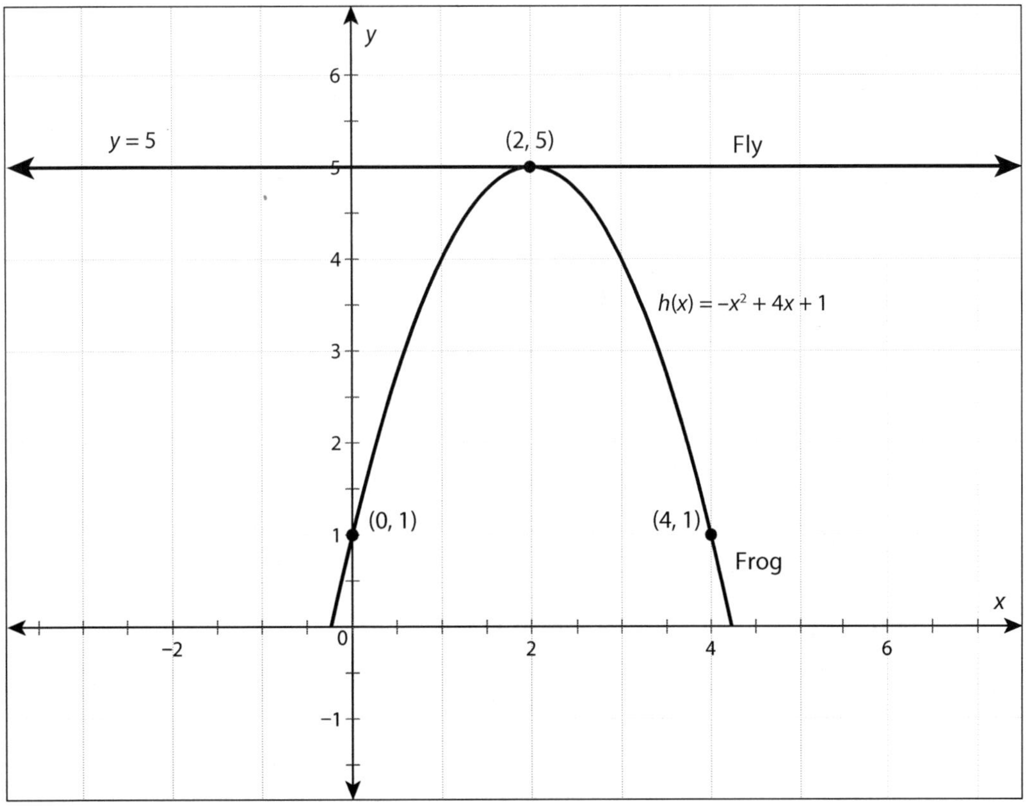

F–IF.7a★; F–IF.8a • Functions

Interpreting Various Forms of Quadratic Functions

Common Core State Standards

F–IF.7

Graph functions expressed symbolically and show key features of the graph, by hand in simple cases and using technology for more complicated cases.★

a. Graph linear and quadratic functions and show intercepts, maxima, and minima.

F–IF.8

Write a function defined by an expression in different but equivalent forms to reveal and explain different properties of the function.

a. Use the process of factoring and completing the square in a quadratic function to show zeros, extreme values, and symmetry of the graph, and interpret these in terms of a context.

Problem-Based Task: Is the Glider Safe?

Shin is a beginner hang glider. He's practicing jumping from a certain height, dipping initially, and then rising. Shin should dip to a height no lower than 6 feet above the ground, which is considered a safe height, before changing direction and beginning to rise. The position of Shin's hang glider is given by $y = (x - 4)(x - 6)$, with x representing the time in seconds since Shin starts the initial jump and y representing the distance in feet from the safe height. Will Shin stay above the safe height? How long will it take for Shin to reach the initial height of the jump?

Problem-Based Task: Is the Glider Safe?

Coaching

a. What is the minimum height Shin reaches?

b. Does a negative answer make sense in the context of the problem? Explain.

c. Find the time at which the minimum height occurs.

d. What are the *x*-intercepts and what do they represent?

e. Will Shin stay above the safe height? Explain.

f. How long will it take for Shin to reach his initial height?

Problem-Based Task: Is the Glider Safe?

Coaching Sample Responses

a. What is the minimum height Shin reaches?

The minimum occurs at the vertex. Write the equation in standard form.

$y = (x - 4)(x - 6)$	Original equation
$y = x^2 - 6x - 4x + 24$	Distribute.
$y = x^2 - 10x + 24$	Simplify.

The vertex is of the form $\left(\frac{-b}{2a}, f\left(\frac{-b}{2a}\right)\right)$.

Use the equation $y = x^2 - 10x + 24$ to find the values of a and b in order to find the x-value of the vertex.

$x = \frac{-b}{2a}$	Formula to find the x-coordinate of the vertex of a quadratic equation
$x = \frac{-(-10)}{2(1)}$	Substitute –10 for a and 1 for b.
$x = 5$	Simplify.

The x-coordinate of the vertex is 5.

Substitute 5 into the original equation to find the y-coordinate.

$y = x^2 - 10x + 24$	Original equation
$y = (5)^2 - 10(5) + 24$	Substitute 5 for x.
$y = -1$	Simplify.

The y-coordinate of the vertex is –1.

The vertex is located at (5, –1).

b. Does a negative answer make sense in the context of the problem? Explain.

Yes. The value of –1 means 1 foot below the safe height.

c. Find the time at which the minimum height occurs.

The x-value of the vertex, 5, minimizes the quadratic.

Shin reaches the minimum 5 seconds after the dive.

d. What are the x-intercepts and what do they represent?

The x-intercepts can be identified from the original equation written in factored form.

$x = 4$ and $x = 6$

The x-intercepts occur at 4 and 6.

The x-intercepts represent times when the glider reaches the safe height.

The glider is at the safe height at 4 seconds and 6 seconds.

e. Does Shin stay above the safe height? Explain.

No. Shin reaches the safe height 4 seconds after the jump and again 6 seconds after the jump.

He dips below the safe height between 4 seconds and 6 seconds.

For 2 seconds, the glider is below the safe height.

f. How long will it take for Shin to reach his initial height?

The initial height occurs when $x = 0$.

The axis of symmetry occurs at the vertex, (5, –1).

The axis of symmetry is $x = 5$.

Zero is 5 units to the left of the axis of symmetry.

The y-value is the same for the point 5 units to the right of the axis of symmetry.

At $x = 10$, the height should be the same as the initial height.

Substitute 0 and 10 for x to check.

$y = (x - 4)(x - 6)$	$y = (x - 4)(x - 6)$
$y = [(0) - 4][(0) - 6]$	$y = [(10) - 4][(10) - 6]$
$y = 24$	$y = 24$

Shin will reach his initial height of 24 feet after 10 seconds.

F–IF.7b★ • Functions

Square Root and Cube Root Functions

Common Core State Standard

F–IF.7

Graph functions expressed symbolically and show key features of the graph, by hand in simple cases and using technology for more complicated cases.★

b. Graph square root, cube root, and piecewise-defined functions, including step functions and absolute value functions.

Problem-Based Task: Auto Acceleration

Car manufacturers test how quickly a car is capable of accelerating by recording how long it takes the car to accelerate from 0 miles per hour to 60 miles per hour. Sofasta Motors' newest car, the Accelerator, reached 60 miles per hour in 8.1 seconds, which is an acceleration of 3.3 meters per second^2. Vroomy Vehicles' car, the VV1, reached 60 miles per hour in 9.1 seconds, which is an acceleration of 2.9 meters per second^2. The time t_a, in seconds, it takes the Accelerator to travel m meters can be represented using the function $t_a = 0.778\sqrt{m}$. The time t_v, in seconds, it takes the VV1 to travel m meters can be represented using the function $t_v = 0.830\sqrt{m}$. How do the differences in the graphs of the functions relate to the differences in the algebraic representations? How do these differences relate to how long it takes each car to travel m meters?

Problem-Based Task: Auto Acceleration

Coaching

a. What is a reasonable domain for the function $t_a = 0.778\sqrt{m}$?

b. What is the corresponding range?

c. What are three points on the graph of $t_a = 0.778\sqrt{m}$?

d. Use the domain, range, and three points to sketch the graph of $t_a = 0.778\sqrt{m}$.

e. What is a reasonable domain for the function $t_v = 0.830\sqrt{m}$?

f. What is the corresponding range?

g. What are three points on the graph of $t_v = 0.830\sqrt{m}$?

h. Use the domain, range, and three points to create the graph of $t_v = 0.830\sqrt{m}$ on the same coordinate plane as the graph of t_a.

i. How are the two graphs of the functions similar?

j. How are the two graphs of the functions different?

k. How are the algebraic representations of each function related to the similarities and differences between the two graphs?

l. What do these differences show about the time it takes the two cars to travel the same distance?

Problem-Based Task: Auto Acceleration

Coaching Sample Responses

a. What is a reasonable domain for the function $t_a = 0.778\sqrt{m}$?

The square root of a negative number is not a real number, so the function is only defined for $m \geq 0$. Because both cars were tested in a short time span of less than 10 seconds, the domain can be restricted to values of m less than or equal to 10.

A reasonable domain is $0 \leq m \leq 10$.

b. What is the corresponding range?

A square root function is either always increasing or always decreasing; the minimum and maximum values of the range will be at the minimum and maximum values of the domain.

For $m = 0$:

$t_a = 0.778\sqrt{0}$

$t_a = 0.778(0)$

$t_a = 0$

For $m = 10$:

$t_a = 0.778\sqrt{10}$

$t_a \approx 0.778(3.162)$

$t_a \approx 2.46$

The corresponding range is $0 \leq t_a \leq 2.46$.

c. What are three points on the graph of $t_a = 0.778\sqrt{m}$?

The minimum and maximum points, at the beginning and end of the domain and range, are two critical points on the graph.

Minimum: (0, 0)

Maximum: (10, 2.46)

Evaluate the function at a value of m between 0 and 10. A square value of m, such as 4, will be easier to evaluate.

Let $m = 4$.

$t_a = 0.778\sqrt{4}$

$t_a = 0.778(2)$

$t_a = 1.556$

A third point on the function is (4, 1.556).

d. Use the domain, range, and three points to create the graph of $t_a = 0.778\sqrt{m}$.

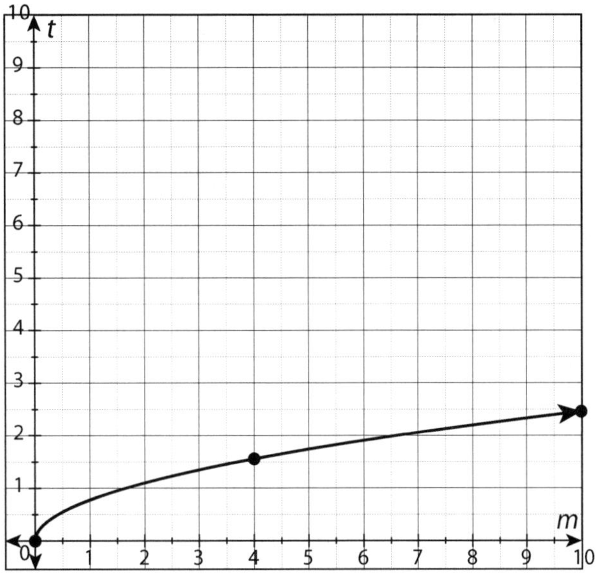

e. What is a reasonable domain for the function $t_v = 0.830\sqrt{m}$?

The square root of a negative number is not a real number, so the function is only defined for $m \geq 0$. Use the same domain restriction as the function t_a: $0 \leq m \leq 10$.

f. What is the corresponding range?

A square root function is either always increasing or always decreasing; the minimum and maximum values of the range will be at the minimum and maximum values of the range.

For *m* = 0:

$t_v = 0.830\sqrt{0}$

$t_v = 0.8330(0)$

$t_v = 0$

For *m* = 10:

$t_v = 0.830\sqrt{10}$

$t_v \approx 0.8330(3.162)$

$t_v \approx 2.62$

The corresponding range is $0 \leq t_a \leq 2.62$.

g. What are three points on the graph of $t_v = 0.830\sqrt{m}$?

The minimum and maximum points, at the beginning and end of the domain and range, are two critical points on the graph.

Minimum: (0, 0)

Maximum: (10, 2.62)

Evaluate the function at a value of *m* between 0 and 10. A square value of *m*, such as 4, will be easier to evaluate.

Let $m = 4$.

$t_v = 0.830\sqrt{4}$

$t_v \approx 0.830(2)$

$t_v \approx 1.66$

A third point on the function is (4, 1.66).

h. Use the domain, range, and three points to create the graph of $t_v = 0.830\sqrt{m}$ on the same coordinate plane as the graph of t_a.

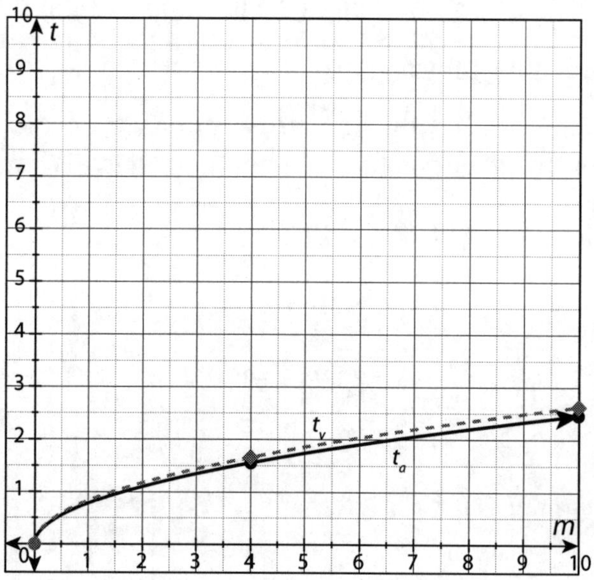

i. How are the two graphs of the functions similar?

Both graphs have the same general shape and same minimum at (0, 0).

j. How are the two graphs of the functions different?

Except for the point (0, 0), the graph of t_v is always greater than the graph of t_a, and the distance between the two graphs is increasing as x gets larger.

k. How are the algebraic representations of each function related to the similarities and differences between the two graphs?

The function t_v has a greater coefficient of the square root, and this graph is always above the graph of the function t_a. Neither function has any addends, either in the square root or outside it, which is why both graphs have a minimum at (0, 0). Both graphs are increasing, which is due to the positive coefficient of the square root.

l. What do these differences show about the time it takes the two cars to travel the same distance?

The outputs of both functions are time to travel a given distance. The car that has the greater t-values takes longer to travel a given distance. If both cars accelerate following the functions t_a and t_v, and start accelerating at the same time, the Accelerator will always be behind the VV1.

F–IF.7b★ • Functions

Absolute Value and Step Functions

Common Core State Standard

F–IF.7

Graph functions expressed symbolically and show key features of the graph, by hand in simple cases and using technology for more complicated cases.★

b. Graph square root, cube root, and piecewise-defined functions, including step functions and absolute value functions.

Problem-Based Task: Bike Rentals

A bike store manager reserves a certain number of bikes to be used as rentals. The manager uses the city's information about tourists to estimate how many bikes to reserve each week; however, she does not want to reserve too many bikes. The manager uses the greatest integer function in her estimate. If the city estimates a total of t tourists between 100 and 1,000 in one week, the bike store manager estimates that $r = \lfloor 0.2t \rfloor + 10$ rentals will be needed. The manager also reserves some tandem bicycles which each fit two riders. The number of tandem bikes reserved, n, is estimated using the function $n = \lfloor 0.10t \rfloor$. How do the differences in the graphs of the functions relate to the differences in the algebraic representations? What do these differences show about the number of regular bikes and tandem bikes reserved?

Problem-Based Task: Bike Rentals

Coaching

a. What is a reasonable domain for the function $r = \lfloor 0.2t \rfloor + 10$?

b. What is the corresponding range?

c. Use a calculator to graph $r = \lfloor 0.2t \rfloor + 10$, and then sketch the graph.

d. What is the domain of the function $n = \lfloor 0.10t \rfloor$ from the problem statement?

e. What is the corresponding range?

f. Graph $n = \lfloor 0.10t \rfloor$ on the calculator on the same coordinate plane as the graph of $r = \lfloor 0.2t \rfloor + 10$. Sketch the graph on the same coordinate plane you used for part c.

g. How are the two graphs of the functions similar?

h. How are the two graphs of the functions different?

i. How are the algebraic representations of each function related to the similarities and differences between the two graphs?

j. What do these differences show about the number of regular bikes reserved and the number of tandem bikes reserved?

Problem-Based Task: Bike Rentals

Coaching Sample Responses

a. What is a reasonable domain for the function $r = \lfloor 0.2t \rfloor + 10$?

The domain is the set of *t* values, which is the number of tourists. The numbers of bike rentals for tourists between 100 and 1,000 are being compared, so the domain is $100 \le t \le 1000$.

b. What is the corresponding range?

The greatest integer function is always increasing or always decreasing. The extreme values of the range will be at the extreme values of the domain. Find the smallest and largest values of the range by evaluating the function at the largest and smallest values of the domain.

For $r = 100$:

$r = \lfloor 0.2(100) \rfloor + 10$

$r = \lfloor 20 \rfloor + 10$

$r = 20 + 10$

$r = 30$

For $r = 1000$:

$r = \lfloor 0.2(1000) \rfloor + 10$

$r = \lfloor 200 \rfloor + 10$

$r = 200 + 10$

$r = 210$

The minimum value of the range is 30, and the maximum value of the range is 210. The greatest integer function has no coefficient and has no addend, so the outputs of the function will be integers.

The range is all integers from 30 to 210.

c. Use a calculator to graph $r = \lfloor 0.2t \rfloor + 10$, and then sketch the graph.

The brackets are the greatest integer function, also called the floor function.

On a TI-83/84:

Step 1: Press [Y=]. If anything is already entered, press [CLEAR].

Step 2: Press [MATH]. Arrow to "NUM," then select 5: int(.

Step 3: Enter the expression contained within the greatest integer function, using *x* as the variable. Enter the remaining part of the expression by moving the cursor to the left and/or right of the int() expression. Type [I][N][T][(][0][.][2][×][X][)][+][1][0].

Step 4: Press [GRAPH].

To change the scale:

Step 5: Press [WINDOW].

Step 6: Set the Xmin to 100, Xmax to 1,000, Xscl to 100, Ymin to 30, Ymax to 210, and Yscl to 20.

On a TI-Nspire:

Step 1: Press the [home] key.

Step 2: Arrow over to the graphing icon and press [enter].

Step 3: Enter the expression at the bottom of the application, using x as the variable. Press the catalog key (to the right of the number 9). Select floor() to enter a greatest integer expression. Type [F][L][O][O][R][(][0][.][2][×][X][)][+][1][0].

Step 4: Press [enter] to plot the expression on the graph.

To change the scale:

Step 5: Press [menu].

Step 6: Select 4: Window.

Step 7: Select 1: Window Settings.

Step 8: Tab between the fields to edit the values. Set the XMin to 100, XMax to 1,000, YMin to 30, and YMax to 210. Leave the XScale and YScale on Auto.

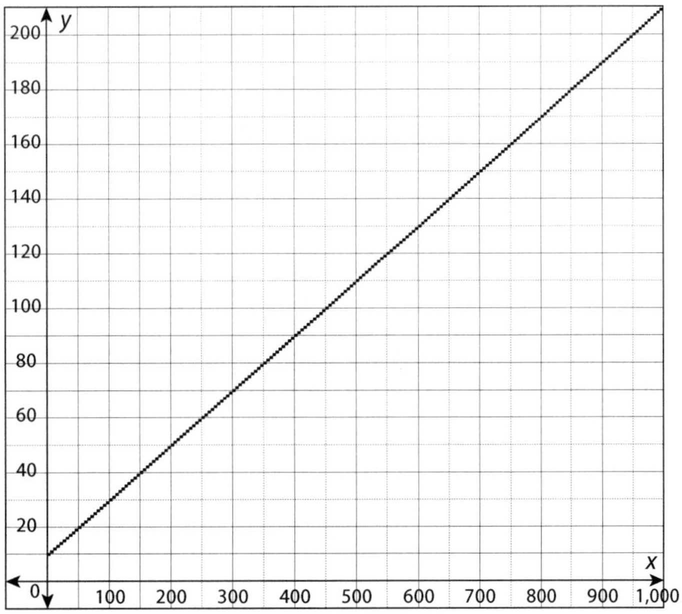

Note: Each step in the graphed function is inclusive (closed circle) at the left boundary and is non-inclusive (open circle) at the right boundary.

d. What is the domain of the function $n = \lfloor 0.10t \rfloor$ from the problem statement?

The domain of $n = \lfloor 0.10t \rfloor$ is the same as the domain for the bike rental function, since both functions have t, the number of tourists, as the input.

$100 \le t \le 1000$

e. What is the corresponding range?

Evaluate the function at the minimum and maximum values of the domain to find the extreme values of the range.

For t = 100:	**For t = 1000:**
$n=\lfloor 0.10(100) \rfloor$	$n=\lfloor 0.10(1000) \rfloor$
$n=\lfloor 10 \rfloor$	$n=\lfloor 100 \rfloor$
$n=10$	$n=100$

The range is all integers from 10 through 100.

f. Graph $n=\lfloor 0.10t \rfloor$ on the calculator on the same coordinate plane as the graph of $r=\lfloor 0.2t \rfloor+10$. Sketch the graph on the same coordinate plane you used for part c.

On the TI-83/84, enter the expression in Y_2. On the TI-Nspire, enter the expression and press enter.

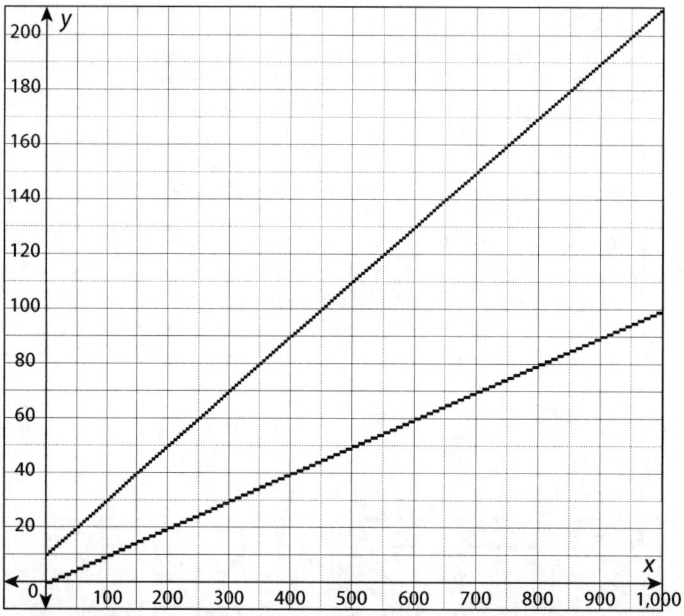

Note: Each step in the graphed functions is inclusive (closed circle) at the left boundary and is non-inclusive (open circle) at the right boundary.

g. How are the two graphs of the functions similar?

Both graphs are step functions, and made up of many disconnected constant functions. Both graphs are increasing.

h. How are the two graphs of the functions different?

The function for the regular bikes is increasing at a faster rate than the function for the tandem bikes. The lengths of the steps on the tandem bike function are also longer. The regular bike function graph has been shifted to the right.

i. How are the algebraic representations of each function related to the similarities and differences between the two graphs?

The coefficient of the variable inside the greatest integer expression changes how quickly the functions increase. The addend outside the greatest integer expression moves the graph of the regular bike estimate to the left.

j. What do these differences show about the number of regular bikes reserved and the number of tandem bikes reserved?

The number of tandem bikes reserved is always less than the number of regular bikes reserved.

NAME:

F–IF.7b★ • Functions

Piecewise Functions

Common Core State Standard

F–IF.7

Graph functions expressed symbolically and show key features of the graph, by hand in simple cases and using technology for more complicated cases.★

b. Graph square root, cube root, and piecewise-defined functions, including step functions and absolute value functions.

Problem-Based Task: Stop and Go

A driver is stopped at a red light. When the light turns green, the driver will accelerate, then drive at a steady speed. The total distance traveled by the car, *d*, in meters, at any time *t*, in seconds, can be represented using the piecewise function $d_1 = \begin{cases} 1.34t^2, \text{ if } 0 \le t \le 10 \\ 26.8t - 134, \text{ if } t > 10 \end{cases}$. If the driver continues to accelerate, the function $d_2 = 1.34t^2$ would represent the driver's distance traveled. What is the difference between the distance traveled if the driver only accelerates for 10 seconds and if the driver continues to accelerate? Use graphs and algebraic representations to explain your answer.

Problem-Based Task: Stop and Go

Coaching

a. What is a reasonable domain to use for both functions?

b. Create a graph of the piecewise function.

c. How are the two pieces of the function different?

d. Create a graph of the quadratic function, d_2.

e. Are there any similarities between the graphs of the two functions?

f. How are these seen in the algebraic equations?

g. Describe the differences between the two functions, including both descriptions of the graphs and algebraic equations.

h. What do these differences and similarities tell you about how far a car would travel following each function?

Problem-Based Task: Stop and Go

Coaching Sample Responses

a. What is a reasonable domain to use for both functions?

The minimum value of the piecewise function is 0. Use a maximum value that is larger than the maximum value of the first piece so that the behavior of the second piece can be observed. The value $t = 30$ is a good maximum of the domain. Therefore, the domain is $0 \le t \le 30$.

b. Create a graph of the piecewise function.

The first part of the function is quadratic. It may help to find three points on this piece of the function to understand its shape. Evaluate the function at three values of t on the domain of the first piece, including the start and end values of the domain.

For $t = 0$:	**For $t = 1$:**	**For $t = 10$:**
$d_1 = 1.34(0^2)$	$d_1 = 1.34(1^2)$	$d_1 = 1.34(10^2)$
$d_1 = 1.34(0)$	$d_1 = 1.34(1)$	$d_1 = 1.34(100)$
$d_1 = 0$	$d_1 = 1.34$	$d_1 = 134$

(0, 0), (1, 1.34), and (10, 134) are all points on the first piece of the function d_1.

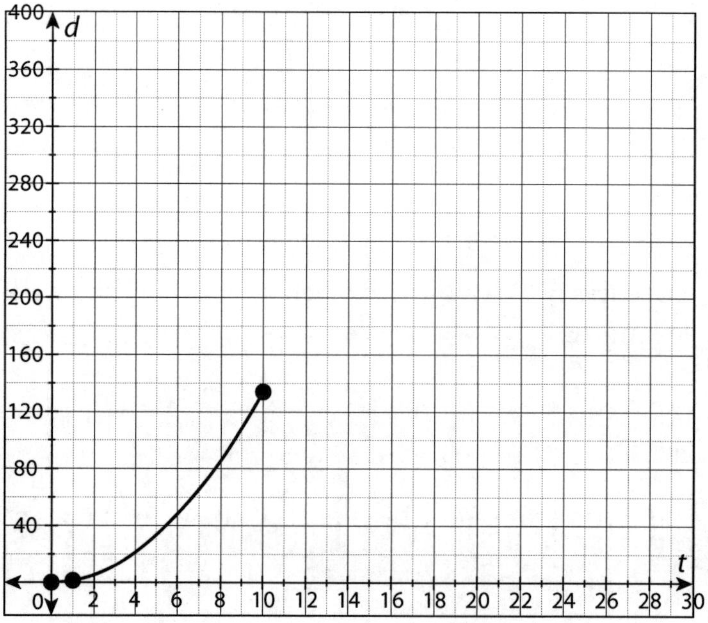

The second piece of the function is linear, so only two points on the line are needed. Find the value of the function at the minimum of the domain, $t = 10$, and the determined maximum, $t = 30$.

For $t = 10$:

$d_1 = 26.8(10) - 134$

$d_1 = 268 - 134$

$d_1 = 134$

For $t = 30$:

$d_1 = 26.8(30) - 134$

$d_1 = 804 - 134$

$d_1 = 670$

(10, 134) is not included on the function, but would be represented by an open point because it is an extreme value of this piece of the function. The point (30, 670) is on the second piece of the function.

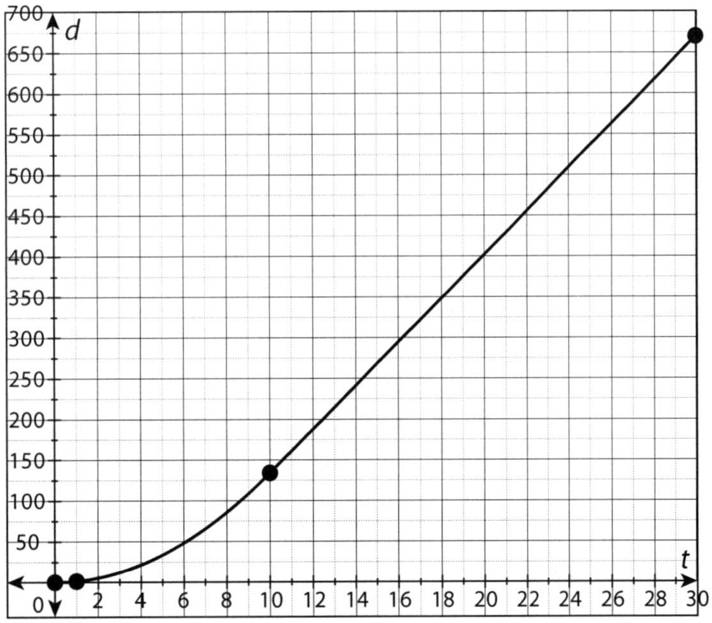

c. How are the two pieces of the function different?

The first piece of the function, the quadratic piece, is curved. The second piece of the function, the linear piece, is a line. The distance is increasing at a constant rate for the second piece of the function. The distance is increasing at a variable rate for the first piece of the function.

d. Create a graph of the quadratic function, d_2.

The graph of d_1 for $0 \le d \le 10$ is the same as the graph of d_2, since they are defined by the same algebraic equation. The first three determined points on d_1 are also on d_2. Find a fourth point on d_2, for $t > 10$, to see the shape of the rest of d_2.

Let $t = 20$.

$d_2 = 1.34(20^2)$

$d_2 = 1.34(400)$

$d_2 = 536$

The point (20, 536) is a fourth point on the graph of d_2.

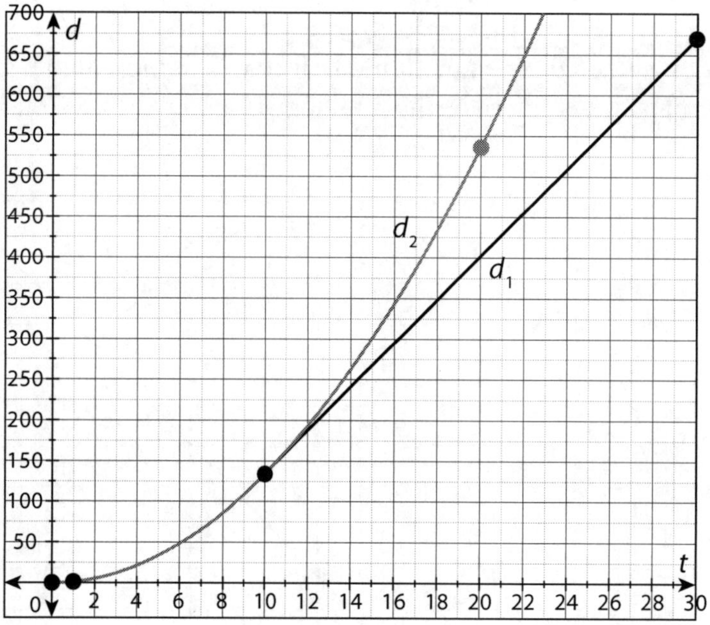

e. Are there any similarities between the graphs of the two functions?

Yes. The graphs are identical for $0 \leq t \leq 10$.

f. How are these seen in the algebraic equations?

The functions for $0 \leq t \leq 10$ are the same for both functions.

g. Describe the differences between the two functions, including both descriptions of the graphs and algebraic equations.

The graphs of the functions for $t > 0$ are both increasing, but the graph of d_2 is increasing faster than the graph of d_1. The graph of d_1 is also increasing at a constant rate for $t > 10$, because the function is linear, where the graph of d_2 is increasing at a changing rate because the function is quadratic.

h. What do these differences and similarities tell you about how far a car would travel following each function?

The output, *d*, is the total distance traveled. For $x > 10$, d_2 is always greater than d_1. This means that a car following the function d_2 will travel farther than a car following function d_1 if both cars began traveling at the same time. The car following function d_2 will also continue to increase its speed, whereas the car following function d_1 would drive at the same speed for $t > 10$ seconds.

NAME:

F–IF.8b • Functions

Analyzing Exponential Functions

Common Core State Standard

F–IF.8

Write a function defined by an expression in different but equivalent forms to reveal and explain different properties of the function.

b. Use the properties of exponents to interpret expressions for exponential functions. *For example, identify percent rate of change in functions such as* $y = (1.02)^t$, $y = (0.97)^t$, $y = (1.01)^{12t}$, $y = (1.2)^{t/10}$, *and classify them as representing exponential growth or decay.*

Problem-Based Task: Which Car Is a Better Buy?

Kristin is buying a new car, and is trying to decide between two particular models. Both cars are the same price today. Kristin knows that the value of each car will change over time. According to her research, the first car, the Roadvana, will have a value of v_1 dollars after t years, modeled by the equation $v_1 = 22{,}000(0.90)^{2t}$. The second car, the Savannah, will have a value of v_2 dollars after t years, modeled by the equation $v_2 = 22{,}000(0.86)^t$. Kristin wants to buy the car that will have a higher value in the future. What is the value of both cars today? Which car should Kristin buy?

Problem-Based Task: Which Car Is a Better Buy?

Coaching

a. Which quantity of each equation represents the value of the car at $t = 0$, or today?

b. Rewrite the equation for the value of the Roadvana with a simplified exponential expression.

c. What is the yearly rate of change in the value of the Roadvana?

d. What is the yearly rate of change in the value of the Savannah?

e. Does each car's value grow or decay over time?

f. Which car will have a higher value in the future?

Problem-Based Task: Which Car Is a Better Buy?

Coaching Sample Responses

a. Which quantity of each equation represents the value of the car at $t = 0$, or today?

In an exponential function of the form $f(t) = a(1 + r)^t$, a is the value of the dependent quantity, $f(t)$, when $t = 0$.

Identify a in each equation to find the value of each car today.

The value of the Roadvana is represented by $v_1 = 22{,}000(0.90)^{2t}$. In this equation, the factor 22,000 represents a. Therefore, the initial value of the Roadvana is $22,000.

The value of the Savannah is represented by $v_2 = 22{,}000(0.86)^t$. The factor 22,000 represents a. Therefore, the initial value of the Savannah is also $22,000.

b. Rewrite the equation for the value of the Roadvana with a simplified exponential expression.

Use the Power of a Power Property to simplify the exponential expression $(0.90)^{2t}$.

$$0.90^{2t} = \left(0.90^2\right)^t = 0.81^t$$

Replace the exponential expression $(0.90)^{2t}$ with the simplified expression in the equation for v_1.

$$v_1 = 22{,}000(0.90)^{2t}$$

$$v_1 = 22{,}000(0.81)^t$$

c. What is the yearly rate of change in the value of the Roadvana?

In an exponential function of the form $f(t) = a(1 + r)^t$, r is the rate of change after each year t. To find the rate of change, subtract 1 from the base of the exponent.

$$v_1 = 22{,}000(0.81)^t$$

$$r = 0.81 - 1 = -0.19$$

The yearly rate of change for the Roadvana is –0.19, or a decrease of 19%.

d. What is the yearly rate of change in the value of the Savannah?

To find the rate of change, r, subtract 1 from the base of the exponent.

$$v_2 = 22{,}000(0.86)^t$$

$$r = 0.86 - 1 = -0.14$$

The yearly rate of change for the Savannah is –0.14, or a decrease of 14%.

e. Does each car's value grow or decay over time?

The yearly rate of change is a decrease for both cars. The values of both cars decay over time.

f. Which car will have a higher value in the future?

The car with the higher future value is the car with a lower yearly rate of decrease. The Savannah has a lower decrease rate of 14%; therefore, it will have a higher value in the future.

F–BF.1a★ • Functions

Building Functions from Context

Common Core State Standard

F–BF.1

Write a function that describes a relationship between two quantities.★

a. Determine an explicit expression, a recursive process, or steps for calculation from a context.

Problem-Based Task: To Drill or Not to Drill?

Cal owns and operates a small oil field in Texas. The field has 75 oil wells, and each well produces 945 barrels of oil per day. There is enough land in the oil field for Cal to drill more wells, but every additional well will cause oil production to drop by 3 barrels per day for each well. Cal believes that it would be profitable to build more wells in the field, but he is not sure how many to build. Building too few wells won't result in the most possible profit. Building too many wells will not be cost effective and will actually cut his profits. Help Cal create a model that allows him to predict the effect that more wells will have on oil production. Use this model to determine the maximum number of wells Cal should have on his land.

Problem-Based Task: To Drill or Not to Drill?

Coaching

a. Write an expression to describe adding an unknown number of wells.

b. Write an expression to describe the effect of drilling more wells on the number of barrels of oil each well produces per day.

c. What is the function that describes oil production in relation to the number of wells?

d. How many barrels of oil does the field produce with 75 wells?

e. What is the maximum number of barrels that could be produced if wells were added?

f. What number of wells will produce the maximum amount of oil?

g. What is the maximum number of barrels the oil field can produce?

Problem-Based Task: To Drill or Not to Drill?

Coaching Sample Responses

a. Write an expression to describe adding an unknown number of wells.

There are 75 wells already. Let x be the number of oil wells Cal wants to add.

The expression $75 + x$ represents the number of existing wells plus the unknown number of new wells.

b. Write an expression to describe the effect of drilling more wells on the number of barrels of oil each well produces per day.

Currently, each well produces 945 barrels of oil per day.

For each well added, 3 fewer barrels of oil will be produced by each well.

Subtract 3 times the unknown number of new wells (x) from the current output per individual well.

$945 - 3x$

c. What is the function that describes oil production in relation to the number of wells?

The number of barrels produced is the product of the number of wells and the number of barrels of oil that can be produced.

$f(x) = (75 + x)(945 - 3x)$

$f(x) = 70{,}875 + 720x - 3x^2$

Rearranging the terms in order of descending powers yields the following function.

$f(x) = -3x^2 + 720x + 70{,}875$

d. How many barrels of oil does the field produce with 75 wells?

Since there are 75 wells already, this means that there are 0 wells added. Evaluate the oil production function at $f(0)$.

$f(x) = -3x^2 + 720x + 70{,}875$

$f(0) = -3(0)^2 + 720(0) + 70{,}875$

$f(0) = 70{,}875$

The oil field produces 70,875 barrels of oil when it has 75 wells.

e. What is the maximum number of barrels that could be produced?

The coefficient for the x^2 term is negative, so the graph of $f(x)$ will be a downward-facing parabola. Therefore, the vertex of $f(x)$ will be the maximum value of the curve. Identify a, b, and c in the equation of the function.

$f(x) = -3x^2 + 720x + 70{,}875$

Thus, $a = -3$, $b = 720$, and $c = 70{,}875$.

The x-coordinate for the vertex of a parabola is $\frac{-b}{2a}$. Substitute values for a and b into this expression to find x.

$$x = \frac{-b}{2a} = \frac{-(720)}{2(-3)} = \frac{-720}{-6} = 120$$

The y-coordinate of the vertex of $f(x)$ can be determined by substituting 120 for x in the function to find $f(120)$.

$f(x) = -3x^2 + 720x + 70{,}875$

$f(120) = -3(120)^2 + 720(120) + 70{,}875$

$f(120) = 114{,}075$

The vertex of $f(x)$ is (120, 114,075).

f. What number of wells will produce the maximum amount of oil?

Interpret the data: x is the number of wells added. The x-value of the vertex is 120. Therefore, 120 is the maximum number of added wells. There are already 75 wells in the field. This means the maximum number of wells is 75 + 120 = 195 wells.

g. What is the maximum number of barrels the oil field can produce?

Interpret the data: y represents the barrels of oil produced each day, so 114,075 is the maximum number of barrels the oil field can produce each day.

F–BF.1b★ • Functions
Operating on Functions

Common Core State Standard

F–BF.1

Write a function that describes a relationship between two quantities.★

b. Combine standard function types using arithmetic operations. *For example, build a function that models the temperature of a cooling body by adding a constant function to a decaying exponential, and relate these functions to the model.*

Problem-Based Task: Pushing Envelopes

A stationery company makes note cards and matching envelopes in various sizes. Regardless of the size of the note card, the ratio of the height to the width is always 3 : 5. Every envelope is made to be precisely 2 mm wider and 3 mm taller than its matching note card. Finally, the amount of paper required to construct each envelope is equal to 3 times the area of the front of the envelope. The company needs to develop a function to calculate the amount of paper needed for each note card and matching envelope for any possible size combinations.

Problem-Based Task: Pushing Envelopes

Coaching

a. What are the equations for the length and width of any size note card?

b. What is the equation that describes the area of any size note card and thus the amount of paper needed?

c. What are the similar equations for the height and width of the corresponding envelope?

d. Using the equations created up to this point, what is the equation that describes the amount of paper needed for an envelope for a given note card?

e. Build an equation that describes the total amount of paper needed for each note card and envelope pair.

Problem-Based Task: Pushing Envelopes

Coaching Sample Responses

a. What are the equations for the length and width of any size note card?

Each note card the company makes has a height-to-width ratio of 3 : 5. Therefore, the height of each note card can be represented as $3x$, where x is an arbitrary length. Similarly, the width of each note card can be represented as $5x$.

$h(x) = 3x$

$w(x) = 5x$

b. What is the equation that describes the area of any size note card and thus the amount of paper needed?

The area of each note card can be determined by multiplying the height by the width.

$(h \bullet w)(x) = 3x \bullet 5x = 15x^2$

c. What are the similar equations for the height and width of the corresponding envelope?

The height of each envelope is 3 mm taller than each note card. Therefore, the height of each envelope is $h(x) + 3 = 3x + 3$.

The width of each envelope is 2 mm wider than each note card. Therefore, the width of each envelope is $w(x) + 2 = 5x + 2$.

d. Using the equations created up to this point, what is the equation that describes the amount of paper needed for an envelope for a given note card?

The amount of paper required to construct an envelope is equal to 3 times the area of the front of the envelope The area of the front of each envelope can be determined by multiplying the height and width.

$(h(x) + 3)(w(x) + 2) = (3x + 3)(5x + 2) = 15x^2 + 21x + 6$

Multiply this amount by 3 to determine the amount of paper needed for each envelope.

$3(15x^2 + 21x + 6) = 45x^2 + 63x + 18$

e. Build an equation that describes the total amount of paper needed for each note card and envelope pair.

The amount of paper needed for one note card is $(h \bullet w)(x)$.

The amount of paper needed for one envelope is $3[(h(x) + 3)(w(x) + 2)]$.

Add the two functions to find the total amount of paper needed for one note card and envelope pair. Let $j(x)$ represent the total amount of paper for the note card and envelope combined.

$j(x) = (h \bullet w)(x) + 3[(h(x) + 3)(w(x) + 2)]$

$j(x) = (3x)(5x) + 3[(3x + 3)(5x + 2)]$

$j(x) = 15x^2 + 3(15x^2 + 21x + 6)$

$j(x) = 15x^2 + 45x^2 + 63x + 18$

$j(x) = 60x^2 + 63x + 18$

F–BF.3 • Functions

Replacing $f(x)$ with $f(x) + k$ and $f(x + k)$

Common Core State Standard

F–BF.3

Identify the effect on the graph of replacing $f(x)$ by $f(x) + k$, $k f(x)$, $f(kx)$, and $f(x + k)$ for specific values of k (both positive and negative); find the value of k given the graphs. Experiment with cases and illustrate an explanation of the effects on the graph using technology. *Include recognizing even and odd functions from their graphs and algebraic expressions for them.*

Problem-Based Task: The Catch

On the last play of a football game, the offense is on the opposing team's 35-yard line. The offense is losing by 4 points, but can win by making a touchdown. The quarterback backs away 5 yards from behind the line of scrimmage and throws the ball to his receiver, who makes the catch at the goal line for the touchdown and the win. The quarterback's release point is 6 feet above the ground, the same height at which the receiver caught the ball. Also, the ball was thrown such that its maximum height was 15 feet above the ground.

Given the above information, build and graph the equation of the football's path, with the x-axis representing the distance from the line of scrimmage and the y-axis as the height of the football above the ground.

Problem-Based Task: The Catch

Coaching

a. What is the total horizontal distance the ball was thrown?

b. What is the height differential between the release of the ball and its maximum height?

c. What are the x-intercepts of the equation describing the path of the ball, disregarding the point of release and allowing the parabola to begin at the origin, (0, 0)?

d. What is the value of a that gives the parabola the correct arch?

e. Using the value of a and the vertex from part d, write the equation of the path of the ball in vertex form.

f. Translate the equation to account for the line of scrimmage.

g. Translate the equation to account for the height of the point of release and catch.

h. Graph the equation.

Problem-Based Task: The Catch

Coaching Sample Responses

a. What is the total horizontal distance the ball was thrown?

The ball was thrown from 5 yards behind the 35-yard line to the goal line. The ball traveled 40 yards.

b. What is the height differential between the release of the ball and its maximum height?

The ball reached a height of 15 feet, or 5 yards, and was thrown from a height of 6 feet, or 2 yards. Therefore, the height differential between the release of the ball and the maximum height is 3 yards.

c. What are the x-intercepts of the equation describing the path of the ball, disregarding the point of release and allowing the parabola to begin at the origin, (0, 0)?

If the path of the ball started at the origin (0, 0), then $x = 0$ is one of the x-intercepts. If the ball was caught 40 yards later, then it was caught at the coordinates (40, 0), making $x = 40$ the other x-intercept of the equation.

d. What value of a gives the parabola the correct arch?

The vertex of the path of the ball must happen at midway between the x-intercepts $(40 - 0 = 20)$, and from part b we know the height differential was 3 yards. Therefore, the vertex of the equation is (20, 3).

Using vertex form, one of the known coordinates (0, 0), and the vertex, we can establish the value of a for the path of the ball.

$$y = a(x - 20)^2 + 3$$

$$0 = a(0 - 20)^2 + 3$$

$$-3 = a(400)$$

$$-\frac{3}{400} = a$$

e. Using the value of a and the vertex from part d, write the equation of the path of the ball in vertex form.

$$f(x) = -\frac{3}{400}(x - 20)^2 + 3$$

f. Translate the equation to account for the release point.

$$f(x+5) = -\frac{3}{400}(x+5-20)^2 + 3 = -\frac{3}{400}(x-15)^2 + 3$$

g. Translate the equation to account for the height of the point of release and catch.

$$f(x+5)+2 = \left[-\frac{3}{400}(x-15)^2 + 3\right] + 2 = -\frac{3}{400}(x-15)^2 + 5$$

h. Graph the function.

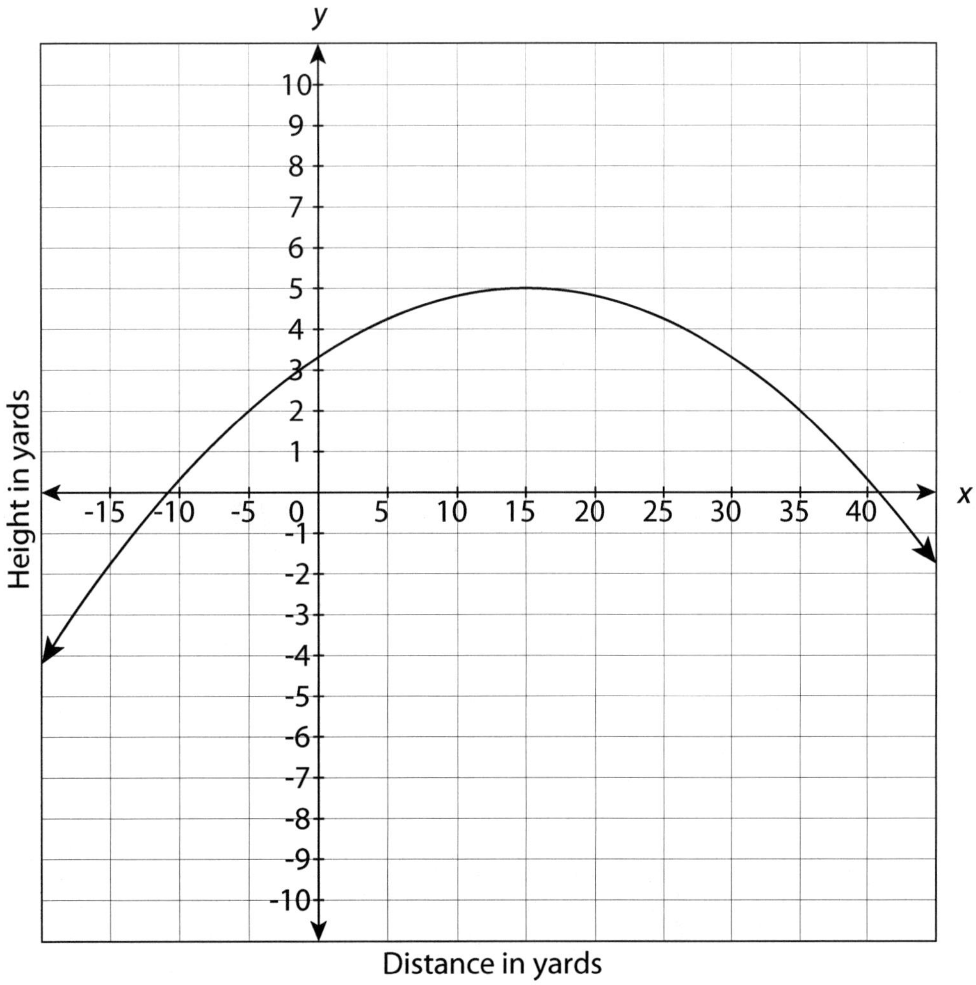

F–BF.3 • Functions

Replacing $f(x)$ with $k \cdot f(x)$ and $f(k \cdot x)$

Common Core State Standard

F–BF.3

Identify the effect on the graph of replacing $f(x)$ by $f(x) + k$, $k f(x)$, $f(kx)$, and $f(x + k)$ for specific values of k (both positive and negative); find the value of k given the graphs. Experiment with cases and illustrate an explanation of the effects on the graph using technology. *Include recognizing even and odd functions from their graphs and algebraic expressions for them.*

Problem-Based Task: Fewer Parabolas, Please

A city hired a civil engineering firm to draw up plans for a 420-foot bridge that would use 7 downward-facing parabolic arches to support the span. The resulting plans called for the arches to be 75 feet high, with a distance of 60 feet between the bases of each arch. A diagram of this plan is shown below.

After seeing the drawings, city councilors asked the civil engineering firm to make a second set of drawings using only 5 evenly spaced parabolic curves but covering the same 420-foot span. What are the equations of the 5 parabolas in the new plan?

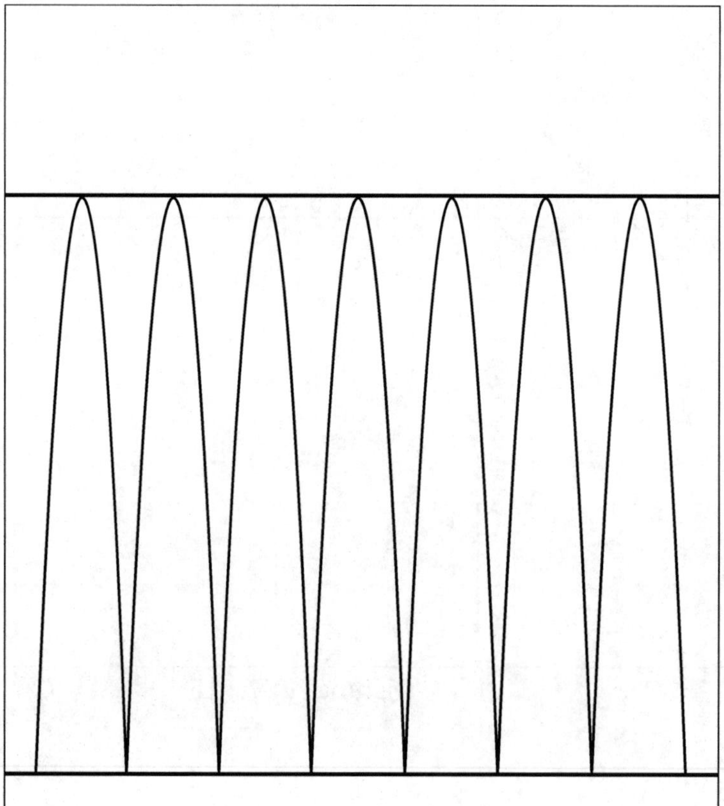

Problem-Based Task: Fewer Parabolas, Please

Coaching

a. Identify the critical points (the vertex and x-intercepts) for the first of the original 7 parabolas.

b. Build the equation in vertex form using the critical points.

c. Solve for a using a point on the parabola.

d. Write the equation that describes each of the 7 parabolas.

e. Determine whether keeping the same height and span of the bridge while reducing the number of arches is an $k \bullet f(x)$ problem or an $f(k \bullet x)$ problem.

f. What is the value of k?

continued

g. What is the first transformed equation?

h. What is the distance between the bases of each arch if the span must be 420 feet?

i. What type of transformation problem must be used to create the remaining four arches?

j. Use the information from parts h and i to create the equations for the four remaining arches.

Problem-Based Task: Fewer Parabolas, Please

Coaching Sample Responses

a. Identify the critical points (the vertex and x-intercepts) for the first of the original 7 parabolas.

Let the y-axis be the line of symmetry for the parabola. The x-value of the vertex is 0. The y-value of the vertex is the maximum value of the arch. The arches are to be 75 feet tall, so the vertex is (0, 75).

The other critical points are where $y = 0$. The arches are 60 feet apart at their base. Divide by 2 to determine the distance each base is from the axis of symmetry.

$60 \div 2 = 30$

Since the y-axis is the axis of symmetry, –30 and 30 are the x-coordinates of the x-intercepts.

Therefore, the x-intercepts of the equation are (–30, 0) and (30, 0).

b. Build the equation in vertex form using the critical points.

Substitute the known information to build the vertex form of the equation.

$f(x) = a(x - h)^2 + k$

$f(x) = a(x - 0)^2 + 75$

$f(x) = ax^2 + 75$

c. Solve for a using a point on the parabola.

Substitute one of the x-intercepts into $f(x) = ax^2 + 75$ to solve for a.

$f(x) = ax^2 + 75$

$0 = a(30)^2 + 75$

$-75 = a \bullet 900$

$$-\frac{75}{900} = a = -\frac{1}{12}$$

d. Write the equation that describes each of the 7 parabolas.

$$f(x) = -\frac{1}{12}x^2 + 75$$

e. Determine whether keeping the same height and span of the bridge while reducing the number of arches is an $k \bullet f(x)$ problem or an $f(k \bullet x)$ problem.

The problem is not to change the height of the bridge but instead to change the number of intervals from the base to the top and back down to the base. This is an $f(k \bullet x)$ problem.

f. What is the value of k?

Starting with the current function, consider the ratio of the number of arches in the old and new plans. We now want 5 parabolas instead of the 7 we had before, so our ratio is 5 : 7.

Remember that a value of k less than 1 will increase the interval between the x-intercepts, so we want our value to be $k=\frac{5}{7}$.

g. What is the first transformed equation?

Substitute the value of k into $f(k \bullet x)$.

$$f(k \bullet x)=f\left(\frac{5}{7}x\right)$$

$$f\left(\frac{5}{7}x\right)=-\frac{1}{12}\left(\frac{5}{7}x\right)^2+75$$

$$f\left(\frac{5}{7}x\right)=-\frac{1}{12}\left(\frac{25}{49}\right)x^2+75$$

$$f\left(\frac{5}{7}x\right)=-\frac{25}{588}x^2+75$$

h. What is the distance between the bases of each arch if the span must be 420 feet?

The problem states the span of the 5 parabolas must stay at 420 feet. There are 5 parabolas. Therefore, divide 420 feet by 5.

$$\frac{420}{5}=84$$

The distance between each pair of bases for each arch is 84 feet.

i. What type of transformation problem must be used to create the remaining four arches?

The first arch needs to be translated to the right 4 times. This is an $f(x + k)$ problem.

j. Use the information from parts h and i to create the equations for the four remaining arches.

For each of the remaining 4 parabolas, translate the function to the right by 84 units each, the distance between the arches.

- Equation 2: $f(x)=-\dfrac{25}{588}(x-84)^2+75$
- Equation 3 (translating Equation 2 by 84 units to the right): $f(x)=-\dfrac{25}{588}(x-168)^2+75$
- Equation 4 (translating Equation 3): $f(x)=-\dfrac{25}{588}(x-252)^2+75$
- Equation 5 (translating Equation 4): $f(x)=-\dfrac{25}{588}(x-336)^2+75$

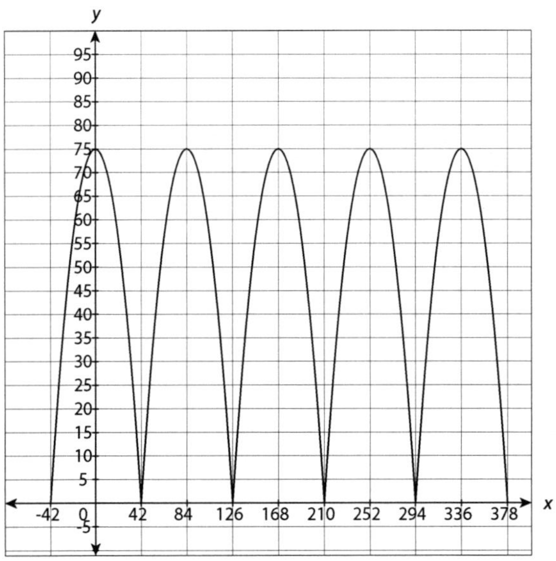

F–BF.4a • Functions

Finding Inverse Functions

Common Core State Standard

F–BF.4

Find inverse functions.

a. Solve an equation of the form $f(x) = c$ for a simple function f that has an inverse and write an expression for the inverse. *For example,* $f(x) = 2x^3$ *or* $f(x) = (x+1)/(x-1)$ *for* $x \neq 1$.

Problem-Based Task: Falling Keys

Piper drops her keys from a third-story window to a friend standing on the sidewalk. The distance between the keys and the ground, in feet, can be represented by the function $f(x) = -16x^2 + 32$, where x is the time in seconds. Piper would like to be able to determine the time, in seconds, at which the keys were any given distance from the ground. What is the function with height in feet from the ground as the independent variable? After about how many seconds were the keys 4 feet above the ground?

Problem-Based Task: Falling Keys

Coaching

a. What type of function is represented by the equation $y = -16x^2 + 32$?

b. Generally, what is the domain of this type of equation?

c. Why does the domain need to be restricted?

d. What is the restricted domain?

e. What is the range of $f(x)$ on the restricted domain?

f. What is the algebraic equation of the function's inverse?

g. Identify the independent and dependent quantities of the function's inverse.

h. What are the domain and range of the function's inverse?

i. After about how many seconds were the keys 4 feet above the ground?

Problem-Based Task: Falling Keys

Coaching Sample Responses

a. What type of function is represented by the equation $y = -16x^2 + 32$?

The equation $y = -16x^2 + 32$ represents a quadratic equation.

b. Generally, what is the domain of this type of equation?

The domain of a quadratic equation is generally all real numbers.

c. Why does the domain need to be restricted?

A quadratic function is not one-to-one; therefore, the inverse of the given function is not a function. Also, since the dependent variable x represents time, the values of x cannot be negative.

d. What is the restricted domain?

A quadratic function of the form $f(x) = a(x - h)^2 + k$ can be restricted to a domain of $x \geq h$ or $x \leq h$ in order for it to be one-to-one. In the function $f(x) = -16x^2 + 32$, $a = -16$, $h = 0$, and $k = 32$. The domain of $f(x)$ can be restricted to $x \geq 0$ or $x \leq 0$. Since the independent variable x represents time, which cannot be negative, the domain of $x \geq 0$ is a better fit for the problem statement.

e. What is the range of $f(x)$ on the restricted domain?

The range of a quadratic function of the form $y = a(x - h)^2 + k$ has an extreme value at $y = k$. If $a > 0$, then the range is $y \geq k$. If $a < 0$, then the range is $y \leq k$. If a quadratic function is restricted to $x \geq h$ or $x \leq h$, then the range is not changed. Since $a < 0$ for the function $f(x) = -16x^2 + 32$, the range is $y \leq 32$.

f. What is the algebraic equation of the function's inverse?

Rewrite the function in the form "$y =$." Switch x and y, and then solve for y.

$f(x) = -16x^2 + 32$

$y = -16x^2 + 32$

$x = -16y^2 + 32$

$16y^2 + x = 32$

$16y^2 = 32 - x$

$$y^2 = \frac{32-x}{16}$$

$$y^2 = 2-\frac{x}{16}$$

$$y=\sqrt{2-\frac{x}{16}}$$

The notation $f^{-1}(x)$ can be used to write the inverse of $f(x)$.

$$f^{-1}(x)=\sqrt{2-\frac{x}{16}}$$

g. Identify the independent and dependent quantities of the function's inverse.

The independent quantity of $f(x)$ was time in seconds, and the dependent quantity of $f(x)$ was distance in feet. The independent and dependent quantities are swapped in the inverse. The independent quantity of $f^{-1}(x)$ is distance in feet, and the dependent quantity of $f^{-1}(x)$ is time in seconds.

h. What are the domain and range of the function's inverse?

The domain of a function's inverse is the range of the function, and the range of a function's inverse is the domain of the function. The domain of $f^{-1}(x)$ is $x \le 32$, and the range of $f^{-1}(x)$ is $y \ge 0$.

i. After about how many seconds were the keys 4 feet above the ground?

Substitute 4 for x in the inverse function, and then solve.

$$f^{-1}(x)=\sqrt{2-\frac{x}{16}}$$

$$f^{-1}(4)=\sqrt{2-\frac{(4)}{16}}$$

$$f^{-1}(4)=\sqrt{2-0.25}$$

$$f^{-1}(4)=\sqrt{1.75}$$

$$f^{-1}(4)\approx 1.32$$

The keys reach a height of 4 feet above the ground after approximately 1.32 seconds.

F–IF.9; F–LE.3★ • Functions

Comparing Properties of Functions Given in Different Forms

Common Core State Standards

F–IF.9

Compare properties of two functions each represented in a different way (algebraically, graphically, numerically in tables, or by verbal descriptions). *For example, given a graph of one quadratic function and an algebraic expression for another, say which has the larger maximum.*

F–LE.3

Observe using graphs and tables that a quantity increasing exponentially eventually exceeds a quantity increasing linearly, quadratically, or (more generally) as a polynomial function.★

Problem-Based Task: On to Washington!

An emergency meeting has been called on Capitol Hill in Washington, D.C. As a newly elected representative for the state of Georgia, you need to find a way to travel to Washington. No flights are available at a time that fits your schedule, but there are three trains that you could take. The distance between Atlanta and Washington, D.C., is 639 miles. Train 1 is a nonstop train that leaves at 10 A.M. and travels at 66 mph. Train 2 also leaves at 10 A.M., but has a few stops that slow it down. The total distance traveled each hour by Train 2 is given in the table below, where t is in hours and $d(t)$ is in miles. Train 3 leaves at noon. The distance traveled for Train 3 can be modeled by the graph below, where t is the time in hours beginning at noon, and d is the distance traveled in miles. Is it possible to determine which train will arrive in Washington first? Which train should you select if you want to minimize time spent on the train? Explain.

t	0	1	2	3	4	5	6	7	8	9	10
$d(t)$	0	48	100	156	216	280	348	420	496	576	660

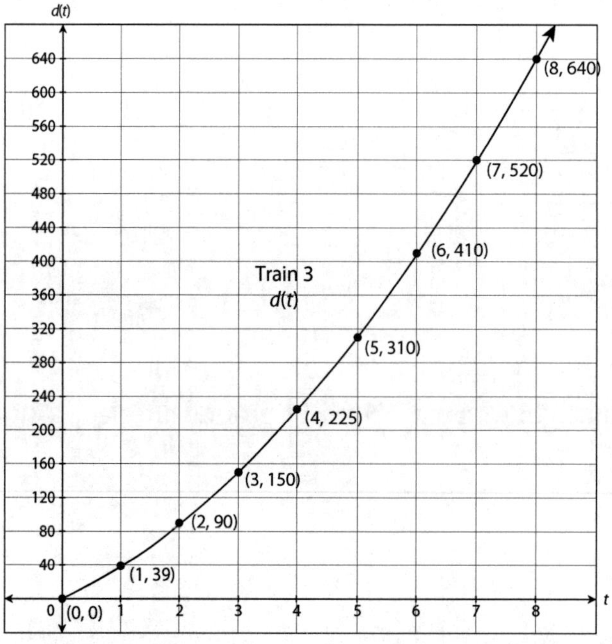

Problem-Based Task: On to Washington!

Coaching

a. Could a table of values help to compare all three trains? If so, create the table.

b. Can all three trains be described by a quadratic model? Why or why not?

c. At what time will Train 1 arrive?

d. At what time will Train 2 arrive?

e. At what time will Train 3 arrive?

f. Is it possible to determine which train will arrive in Washington, D.C., first?

g. Which train should you select if you need to arrive in Washington, D.C., as early as possible? Explain.

h. Which train should you select if you want to minimize the amount of time spent on the train? Explain.

Problem-Based Task: On to Washington!

Coaching Sample Responses

a. Could a table of values help to compare all three trains? If so, create the table.

Yes, a table of values could help to compare the distances traveled in relation to time for all three trains. Use the graph to approximate the distances for Train 3.

Time	10 A.M.	11 A.M.	12 P.M.	1 P.M.	2 P.M.	3 P.M.	4 P.M.	5 P.M.	6 P.M.	7 P.M.	8 P.M.
Train 1	0	66	132	198	264	330	396	462	528	594	660
Train 2	0	48	100	156	216	280	348	420	496	576	660
Train 3	0	0	0	39	90	150	225	310	410	520	640

b. Do all three trains travel in a quadratic model? Why or why not?

No; the first train is linear. It travels at a constant rate of 66 mph. The distance traveled by the second train increases as the hours pass, but not by a constant multiple. Neither the first nor the second differences are constant. The second train is not linear, quadratic, or exponential. The third train does follow a quadratic model.

c. At what time will Train 1 arrive?

Train 1 travels at a constant speed of 66 mph.

To travel 639 miles, it takes $639 \div 66 = 9.68$ hours, or about 9 hours and 41 minutes.

If the train leaves Atlanta at 10 A.M., it will arrive in Washington, D.C., at about 7:41 P.M.

d. At what time will Train 2 arrive?

According to the table, the train will go 576 miles in 9 hours and 660 miles in 10 hours. The distance from Atlanta to Washington is 639 miles, which is between these two points but closer to 10 hours. The train will arrive between 7 P.M. and 8 P.M., but closer to 8 P.M.

e. At what time will Train 3 arrive?

The distance traveled by Train 3 is represented by the graph. To find out when this train will have traveled 639 miles, graph the horizontal line $d(t) = 639$ and observe where the functions intersect. They intersect when x is approximately 8.

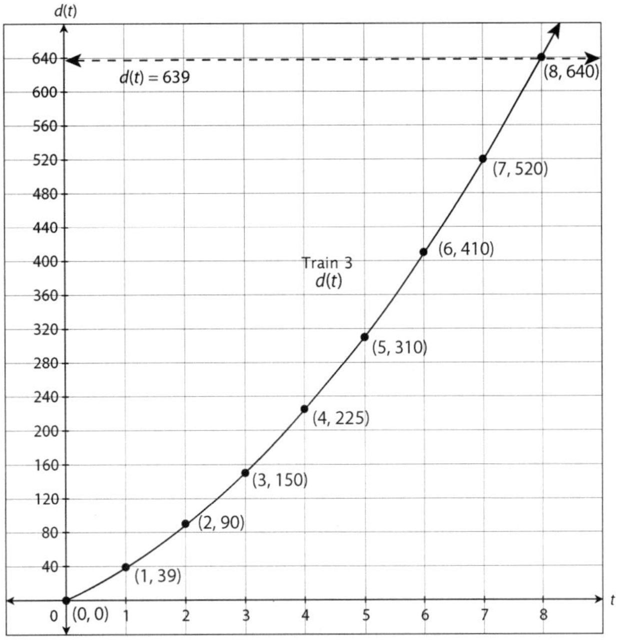

f. Is it possible to determine which train will arrive in Washington, D.C., first?

Train 1 would arrive in Washington, D.C., after 9 hours 41 minutes, or at 7:41 P.M.

It is not possible to determine the exact time for Train 2, but it will arrive between 7 P.M. and 8 P.M.

Using the graph, Train 3 would arrive in Washington, D.C., after approximately 8 hours, or at about 8:00 P.M.

Since an exact arrival time cannot be determined for Train 2, it is not possible to determine which train will arrive in Washington, D.C., first.

g. Which train should you select if you need to arrive in Washington, D.C., as early as possible? Explain.

Because it is not possible to determine which train will arrive in Washington, D.C., first, we cannot say for sure, but we do know that Train 1 arrives about 18 minutes before Train 3.

h. Which train should you select if you want to minimize the amount of time spent on the train? Explain.

Train 3 leaves Atlanta 2 hours later than Trains 1 and 2, but arrives only 18 minutes later than Train 1. You should select Train 3 to minimize the amount of time spent on the train.

F–TF.8 • Functions

Proving the Pythagorean Identity

Common Core State Standard

F–TF.8

Prove the Pythagorean identity $\sin^2(\theta) + \cos^2(\theta) = 1$ and use it to find $\sin(\theta)$, $\cos(\theta)$, or $\tan(\theta)$ given $\sin(\theta)$, $\cos(\theta)$, or $\tan(\theta)$ and the quadrant of the angle.

Problem-Based Task: Slick Surfboards

Sloane makes custom surfboards. He applies surfboard wax to each board to reduce the amount of friction created when the board is in the water, so that the board slides over the water more quickly. Sloane is testing several different surfboard waxes to determine which is best. He tests each wax by first applying it to a surfboard that's lying horizontally. Then, he places a block on the board and slowly lifts one end of the board up at an angle. Sloane's assistant records the angle at which the block begins to slide. The sooner the block slides, the better the wax.

The equation that relates the force of the block to the resistive force of friction is $mg\sin\theta = \mu mg\cos\theta$, where θ = the angle of incline, m = the mass of the block, g = the block's acceleration due to gravity, and μ = the coefficient of friction. The coefficient of friction is a constant that represents the force the object needs to overcome in order to move. If the coefficient of friction, μ, during one test was determined to be 0.12, at what angle did the block begin to slide?

Problem-Based Task: Slick Surfboards

Coaching

a. What is the equation that relates the force of the block to the resistive force of friction?

b. What value(s) are given and can be substituted into the equation?

c. How can you rearrange the equation to isolate the unknowns?

d. What trigonometric identity can you use to simplify the equation?

e. How can you solve for the angle of incline? What is the angle of incline?

Problem-Based Task: Slick Surfboards

Coaching Sample Responses

a. What is the equation that relates the force of the block to the resistive force of friction?

$$mg\sin\theta = \mu mg\cos\theta$$

b. What value(s) are given and can be substituted into the equation?

We are given that the coefficient of friction is 0.12. Therefore, $\mu = 0.12$. No other values are given.

c. How can you rearrange the equation to isolate the unknowns?

First substitute in the known quantity, $\mu = 0.12$. Then, rearrange to solve for θ.

$$mg\sin\theta = \mu mg\cos\theta$$

$$mg\sin\theta = 0.12mg\cos\theta$$

$$\frac{\cancel{mg}\sin\theta}{\cancel{mg}} = \frac{0.12\cancel{mg}\cos\theta}{\cancel{mg}}$$

$$\sin\theta = 0.12\cos\theta$$

$$\frac{\sin\theta}{\cos\theta} = \frac{0.12\cancel{\cos\theta}}{\cancel{\cos\theta}}$$

$$\frac{\sin\theta}{\cos\theta} = 0.12$$

d. What trigonometric identity can you use to simplify the equation?

The tangent ratio identity can be used to replace $\frac{\sin\theta}{\cos\theta}$.

$$\tan\theta = \frac{\sin\theta}{\cos\theta}$$

e. How can you solve for the angle of incline? What is the angle of incline?

After substituting, you are left with the following:

$$\tan\theta = 0.12$$

The ratio is known but the angle is unknown. Use the inverse tangent to find the angle of incline.

$\theta = \tan^{-1}(0.12)$

$\theta \approx 6.8$

The angle of incline is about 7°.

NAME:

G–CO.9 • Geometry

Proving the Vertical Angles Theorem

Common Core State Standard

G–CO.9

Prove theorems about lines and angles. *Theorems include: vertical angles are congruent; when a transversal crosses parallel lines, alternate interior angles are congruent and corresponding angles are congruent; points on a perpendicular bisector of a line segment are exactly those equidistant from the segment's endpoints.*

Problem-Based Task: Cutting Kitchen Tiles

Maresol is retiling the backsplash over her kitchen stove, and has to cut square ceramic tiles into 4 congruent triangles to create the pattern she wants. If she doesn't cut each square into perfectly equal triangles, the tiles won't fit together properly. Before she cuts the first tile, she uses a pencil to draw two segments on the tile. The segments form perpendicular bisectors. Use what you know about triangle congruency and perpendicular bisectors to prove that the 4 triangles are congruent.

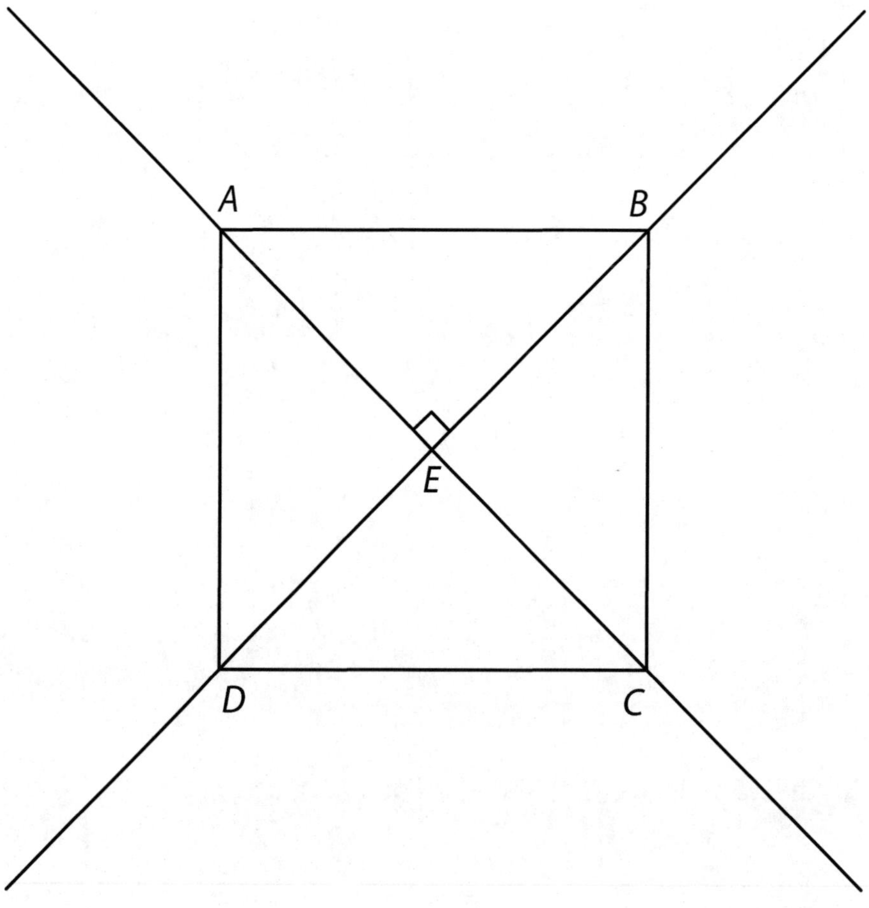

Problem-Based Task: Cutting Kitchen Tiles

Coaching

a. What information are you given?

b. What is the definition of a perpendicular bisector?

c. Use the definition of a perpendicular bisector to determine which segments are congruent. Mark them in your diagram.

d. What does the Perpendicular Bisector Theorem state?

e. Write a statement of congruence using the Perpendicular Bisector Theorem.

f. Go to the next consecutive vertex and write a second statement of congruence using the Perpendicular Bisector Theorem.

g. Go to the next consecutive vertex and write a third statement of congruence using the Perpendicular Bisector Theorem.

h. Go to the next consecutive vertex and write a fourth statement of congruence using the Perpendicular Bisector Theorem.

i. On the diagram, mark the congruent sides that you developed in parts e–h.

j. What criteria do you need to prove triangles are congruent?

k. Are the four sections of the tile congruent?

Problem-Based Task: Cutting Kitchen Tiles

Coaching Sample Responses

a. What information are you given?

$\overline{AC}$ and $\overline{DB}$ are perpendicular bisectors of each other.

b. What is the definition of a perpendicular bisector?

A perpendicular bisector is a line that intersects a segment at a right angle at the midpoint of the segment.

c. Use the definition of a perpendicular bisector to determine which segments are congruent. Mark them in your diagram.

Since $\overline{AC}$ is the perpendicular bisector of $\overline{DB}$, then $\overline{DE} \cong \overline{EB}$.

Since $\overline{DB}$ is the perpendicular bisector of $\overline{AC}$, then $\overline{AE} \cong \overline{EC}$.

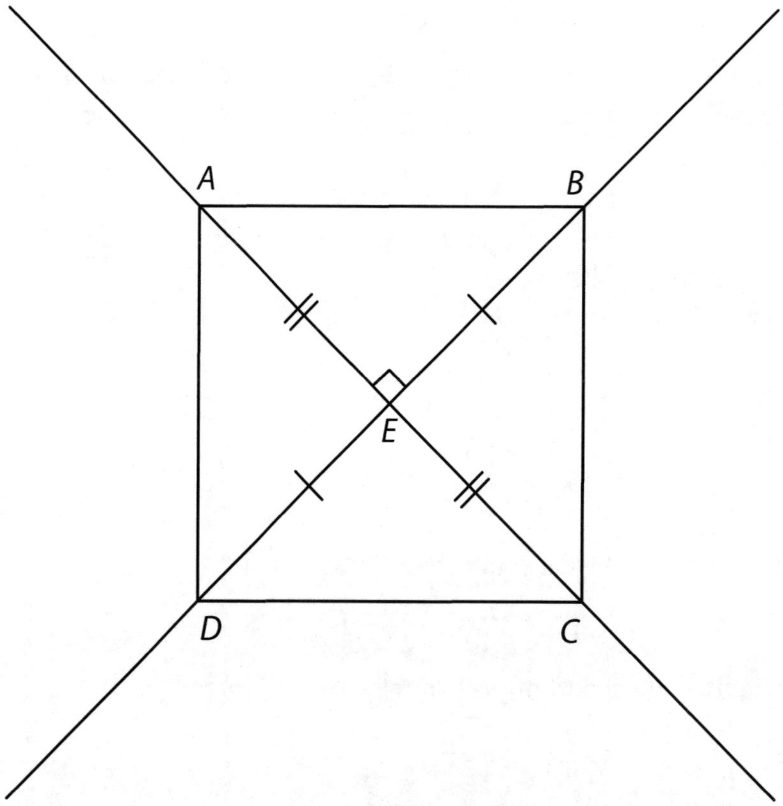

d. What does the Perpendicular Bisector Theorem state?

Any point on a perpendicular bisector is equidistant from the endpoints of the segment it bisects.

e. Write a statement of congruence using the Perpendicular Bisector Theorem.

Since point A is on the perpendicular bisector of $\overline{DB}$, $\overline{AD} \cong \overline{AB}$.

f. Go to the next consecutive vertex and write a second statement of congruence using the Perpendicular Bisector Theorem.

Since point B is on the perpendicular bisector of $\overline{AC}$, $\overline{BA} \cong \overline{BC}$.

g. Go to the next consecutive vertex and write a third statement of congruence using the Perpendicular Bisector Theorem.

Since point C is on the perpendicular bisector of $\overline{DB}$, $\overline{CB} \cong \overline{CD}$.

h. Go to the next consecutive vertex and write a fourth statement of congruence using the Perpendicular Bisector Theorem.

Since point D is on the perpendicular bisector of $\overline{AC}$, $\overline{DC} \cong \overline{DA}$.

i. On the diagram, mark the congruent sides that you developed in parts e–h.

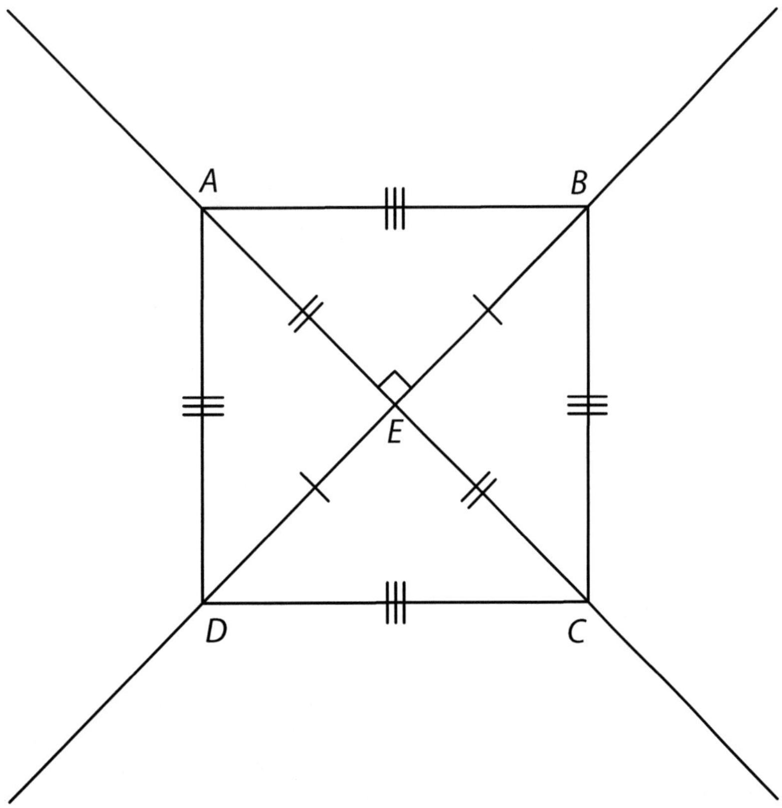

j. What criteria do you need to prove triangles are congruent?

Side-side-side (SSS), side-angle-side (SAS), or angle-side-angle (ASA).

k. Are the four sections of the tile congruent?

Yes, the four triangles are congruent. $\triangle ABE \cong \triangle CBE \cong \triangle CDE \cong \triangle ADE$ by side-side-side (SSS).

G–CO.9 • Geometry

Proving Theorems About Angles in Parallel Lines Cut by a Transversal

Common Core State Standard

G–CO.9

Prove theorems about lines and angles. *Theorems include: vertical angles are congruent; when a transversal crosses parallel lines, alternate interior angles are congruent and corresponding angles are congruent; points on a perpendicular bisector of a line segment are exactly those equidistant from the segment's endpoints.*

Problem-Based Task: Retinal Refraction

When a person looks at an object, the light rays are refracted or distorted as they pass through the eye, and the image is transmitted upside down on the retina in the back of the eye. The object and its retinal image are similar. Prove that they are in proportion using similar triangles. A diagram is given below. Assume that both the person looking at the image and the image are vertical.

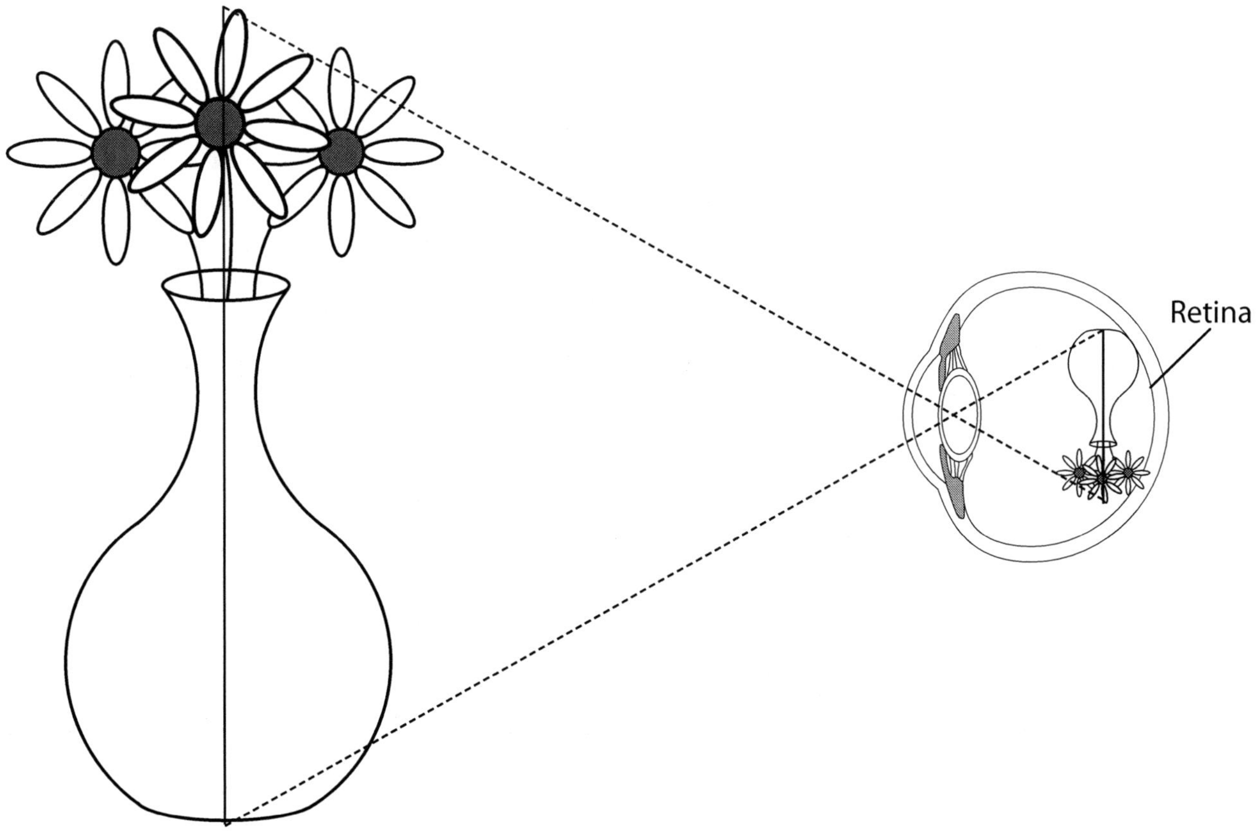

Problem-Based Task: Retinal Refraction

Coaching

a. Label the angles on the diagram.

b. What information is given?

c. Which angles are congruent in parallel lines intersected by a transversal?

d. Which angles are congruent in the diagram and why?

e. What criteria are needed to prove that triangles are similar?

f. Which two triangles are similar?

g. What is the proof that the triangles are similar?

h. What is true about the parts of similar triangles?

Problem-Based Task: Retinal Refraction

Coaching Sample Responses

a. Label the angles on the diagram.

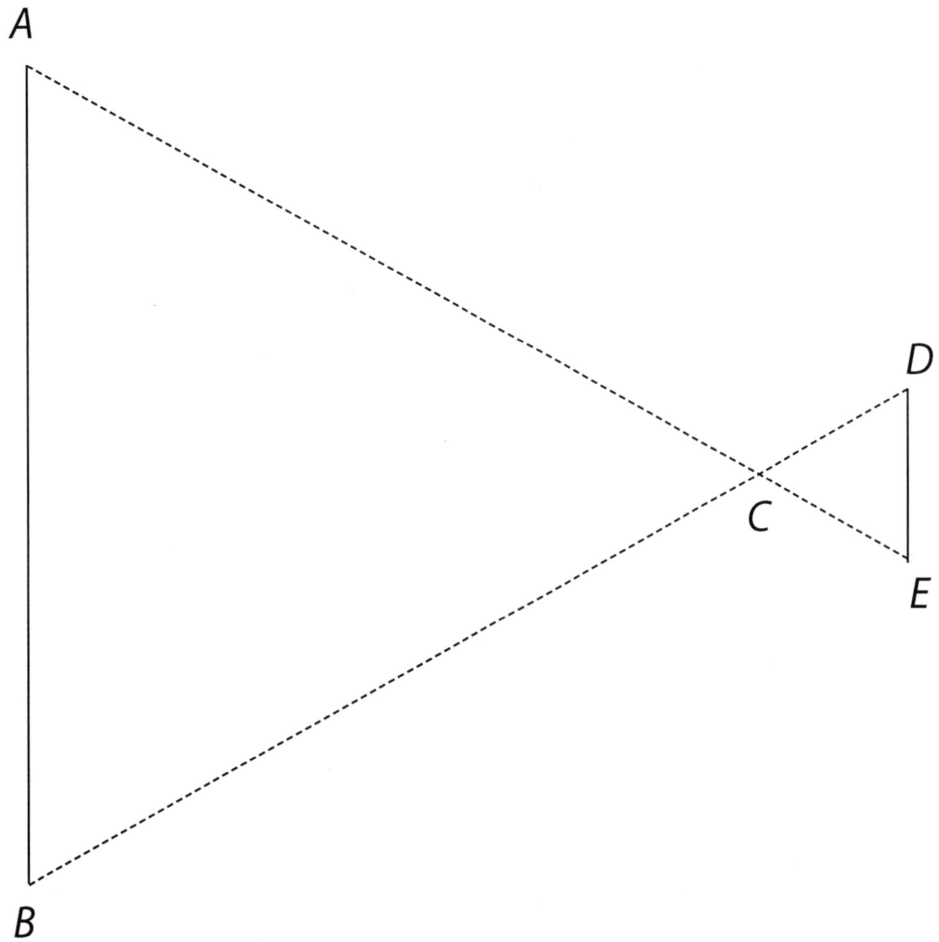

b. What information is given?

$\overline{AB}$ and $\overline{DE}$ are both vertical. Therefore, $\overline{AB} \parallel \overline{DE}$.

c. Which angles are congruent in parallel lines intersected by a transversal?

Alternate interior angles are congruent and alternate exterior angles are congruent.

d. Which angles are congruent in the diagram and why?

$\angle CED \cong \angle BAE$ by the Alternate Interior Angles Theorem.

$\angle CDE \cong \angle ABD$ by the Alternate Interior Angles Theorem.

$\angle ACB \cong \angle DCE$ by the Vertical Angles Theorem.

e. What criteria are needed to prove that triangles are similar?

The needed criteria are two angles (AA), two sides and the included angle (SAS), or all three sides (SSS).

f. Which two triangles are similar?

$\triangle ABC \sim \triangle EDC$

g. What is the proof that the triangles are similar?

Statements	Reasons
1. $\overline{AB}$ and $\overline{DE}$ are both vertical.	1. Given
2. $\overline{AB} \parallel \overline{DE}$	2. Definition of parallel lines
3. $\angle CED \cong \angle BAE$	3. Alternate Interior Angles Theorem
4. $\angle ACB \cong \angle DCE$	4. Vertical Angles Theorem
5. $\triangle ABC \sim \triangle EDC$	5. Angle-Angle Similarity Postulate

h. What is true about the parts of similar triangles?

The corresponding parts of similar triangles are in proportion.

G–CO.10 • Geometry

Proving the Interior Angle Sum Theorem

Common Core State Standard

G–CO.10

Prove theorems about triangles. *Theorems include: measures of interior angles of a triangle sum to 180°; base angles of isosceles triangles are congruent; the segment joining midpoints of two sides of a triangle is parallel to the third side and half the length; the medians of a triangle meet at a point.*

Problem-Based Task: Sensing Distance

Distance-measuring sensors frequently used in robotics send out a beam of infrared light that hits an object. The beam bounces off the object and returns to the sensor's detector, creating a triangle similar to the one below.

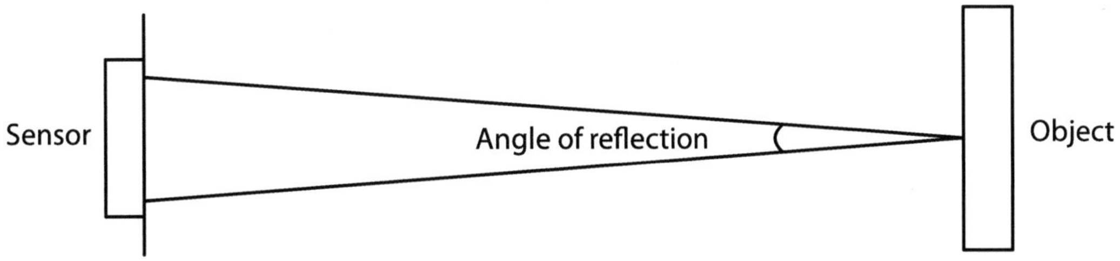

The angles of the triangle vary depending on the sensor's distance from the object. The sensor uses the angles to determine how far away the object is. As the angle of reflection increases, the calculated distance becomes more accurate. In each diagram below, the sensor is parallel to the object. Which of the sensors calculates a more accurate distance: Sensor A, or Sensor B? Explain your reasoning.

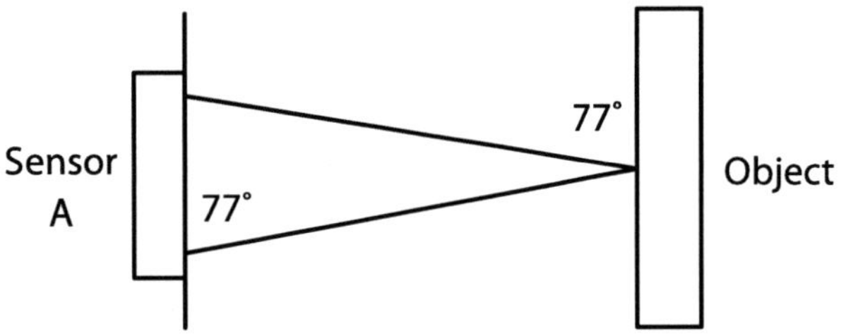

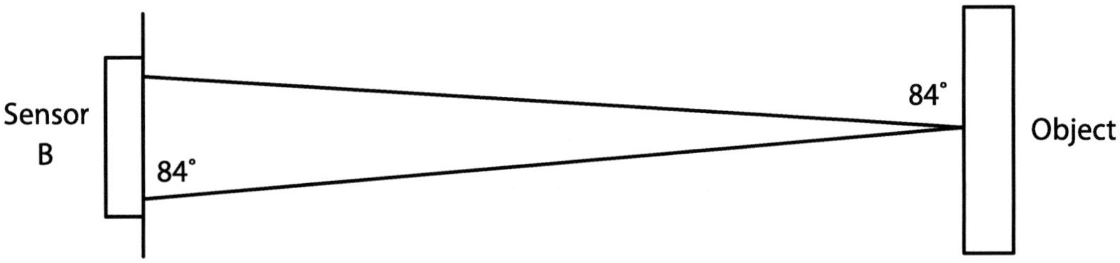

Problem-Based Task: Sensing Distance

Coaching

Use the diagram below to answer the questions that follow.

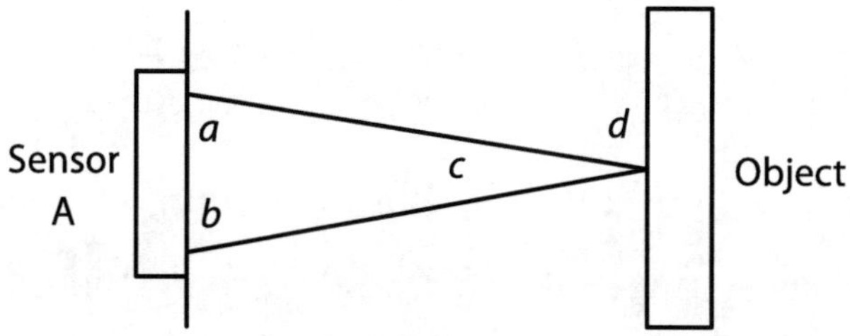

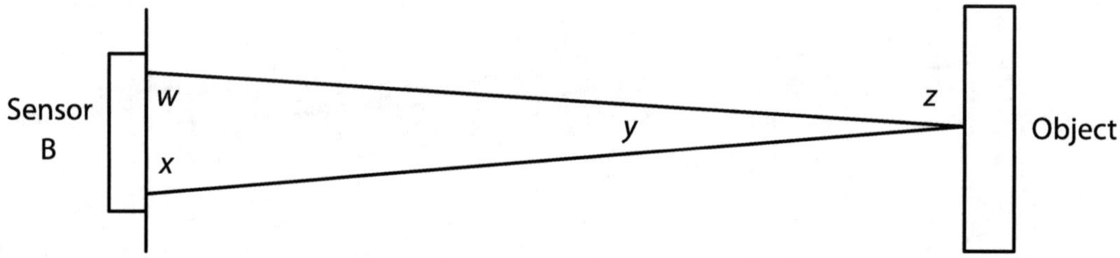

a. What is the relationship between $\angle a$ and $\angle d$?

b. If the measure of $\angle d$ is 77°, what is the measure of $\angle a$?

c. What is the sum of the interior angles of a triangle?

d. What is the measure of $\angle c$?

e. What is the relationship between $\angle z$ and $\angle w$?

f. If the measure of $\angle z$ is 84°, what is the measure of $\angle w$?

g. What is the measure of $\angle y$?

h. Which sensor has a greater angle of reflection?

i. Which sensor is more accurate?

Problem-Based Task: Sensing Distance

Coaching Sample Responses

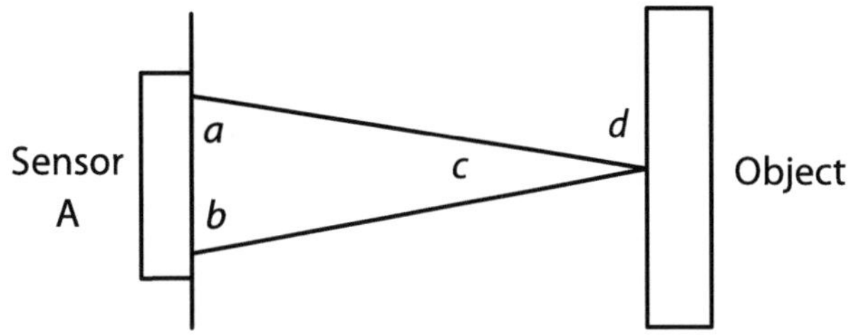

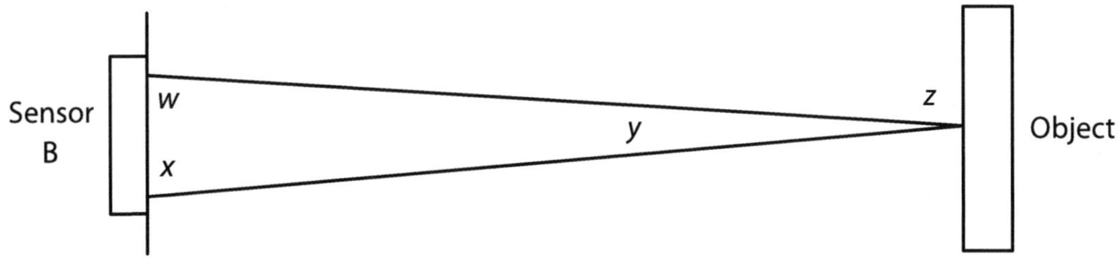

a. What is the relationship between $\angle a$ and $\angle d$?

The sensor and the object are parallel; therefore, the segment from the sensor to the object acts as a transversal.

$\angle a$ and $\angle d$ are alternate interior angles and are congruent.

b. If the measure of $\angle d$ is 77°, what is the measure of $\angle a$?

The angles are congruent; therefore, $\angle a$ is also equal to 77°.

c. What is the sum of the interior angles of a triangle?

The sum of the interior angles of any triangle is 180°.

d. What is the measure of $\angle c$?

To determine the measure of $\angle c$, subtract the sum of $\angle a$ and $\angle b$ from 180.

$$180 - (77 + 77) = 26$$

The measure of $\angle c$ is 26°.

e. What is the relationship between $\angle z$ and $\angle w$?

$\angle z$ and $\angle w$ are alternate interior angles and are congruent.

f. If the measure of $\angle z$ is 84°, what is the measure of $\angle w$?

The angles are congruent; therefore $\angle w$ is also equal to 84°.

g. What is the measure of $\angle y$?

To determine the measure of $\angle y$, subtract the sum of $\angle w$ and $\angle x$ from 180.

Since the triangle is an isosceles triangle, $m\angle x = m\angle w = 84$.

$180 - (84 + 84) = 12$

The measure of $\angle y$ is 12°.

h. Which sensor has a greater angle of reflection?

The angle of reflection for Sensor A is 26°.

The angle of reflection for Sensor B is 12°.

Sensor A has a greater angle of reflection.

i. Which sensor is more accurate?

As the angle of reflection increases, the calculated distance is more accurate.

Sensor A has a greater angle of reflection; therefore, it is more accurate.

G–CO.10 • Geometry

Proving Theorems About Isosceles Triangles

Common Core State Standard

G–CO.10

Prove theorems about triangles. *Theorems include: measures of interior angles of a triangle sum to 180°; base angles of isosceles triangles are congruent; the segment joining midpoints of two sides of a triangle is parallel to the third side and half the length; the medians of a triangle meet at a point.*

Problem-Based Task: Calibrating Consoles

In order for a game console controller to work correctly, users must first calibrate the system. The user stands with the controller in front of the system and allows the system to recognize the controller's location. Optimal performance is achieved when an isosceles triangle, as shown below, is created.

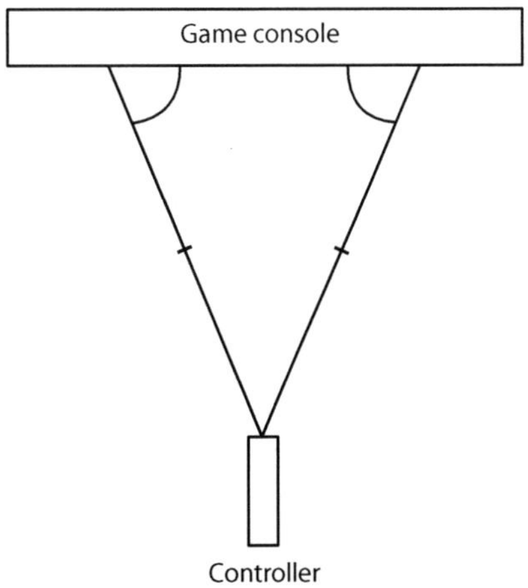

Scarlett and Wren have calibrated their game console controllers as shown in the following diagram. Did either of the girls calibrate her game console controller correctly? Explain your reasoning.

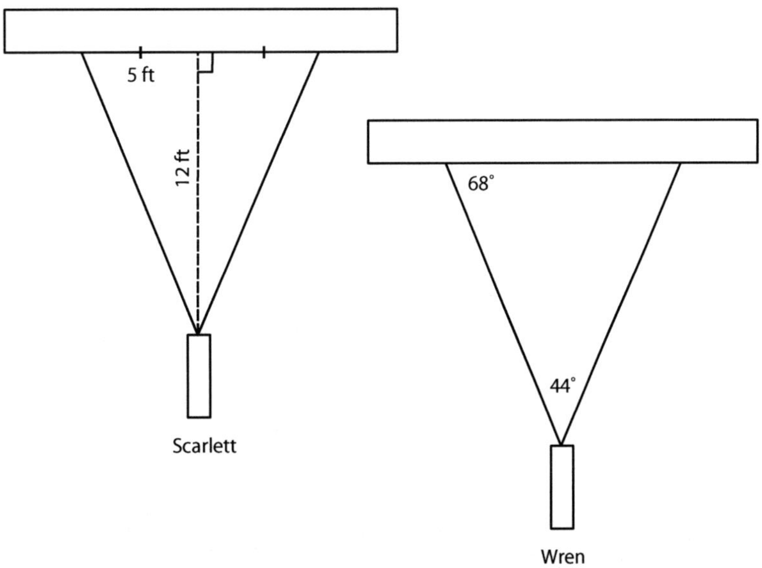

Problem-Based Task: Calibrating Consoles

Coaching

a. What is the definition of an isosceles triangle?

b. How can you determine the length of each side of the triangle Scarlett created?

c. What is the length of each side of the triangle Scarlett created?

d. Is the triangle Scarlett created an isosceles triangle?

e. How can you determine the measure of the third angle of the triangle Wren created?

f. What is the measure of the third angle of the triangle Wren created?

g. Is the triangle Wren created an isosceles triangle?

h. Did either girl calibrate her game console controller correctly? If so, who?

Problem-Based Task: Calibrating Consoles

Coaching Sample Responses

a. What is the definition of an isosceles triangle?

Isosceles triangles have at least two sides that are congruent and at least two base angles that are congruent.

b. How can you determine the length of each side of the triangle Scarlett created?

The triangle Scarlett created is made up of two smaller triangles.

The smaller triangles are congruent because of the Side-Angle-Side Congruence Statement.

Use the Pythagorean Theorem to calculate the lengths of the sides of the triangle Scarlett created.

c. What is the length of each side of the triangle Scarlett created?

$a^2 + b^2 = c^2$	Pythagorean Theorem
$5^2 + 12^2 = c^2$	Substitute 5 and 12 for the lengths of *a* and *b*.
$25 + 144 = c^2$	Simplify.
$169 = c^2$	

$c = \sqrt{169} = 13$

Remember that the two smaller triangles are congruent.

The sides of the triangle Scarlett created are each 13 feet long.

d. Is the triangle Scarlett created an isosceles triangle?

The triangle Scarlett created is an isosceles triangle because two of the sides are equal in length.

e. How can you determine the measure of the third angle of the triangle Wren created?

The sum of the interior angles of a triangle is 180°.

Subtract the given angles from 180 to determine the unknown angle measure.

f. What is the measure of the third angle of the triangle Wren created?

$180 - (68 + 44) = 68$

The third angle measure of the triangle Wren created is 68°.

g. Is the triangle Wren created an isosceles triangle?

The triangle Wren created is an isosceles triangle because two of the angles are congruent.

h. Did either girl calibrate her game console controller correctly? If so, who?

Both girls calibrated their game console controllers correctly because the triangle created by each girl is an isosceles triangle.

G–CO.10 • Geometry

Proving the Midsegment of a Triangle

Common Core State Standard

G–CO.10

Prove theorems about triangles. *Theorems include: measures of interior angles of a triangle sum to 180°; base angles of isosceles triangles are congruent; the segment joining midpoints of two sides of a triangle is parallel to the third side and half the length; the medians of a triangle meet at a point.*

Problem-Based Task: Life-Size Support

Alex is assembling a life-size cardboard cutout of his favorite musician. The package contains three pieces of cardboard: the cutout, which measures 66 inches; the horizontal support, which measures 15.5 inches; and the angled support, which attaches to the back of the cutout at a point that is $\frac{2}{3}$ the length of the cutout from the floor. The horizontal support acts as the midsegment connecting the cutout and the angled support. What is the length of the angled support rounded to the nearest inch?

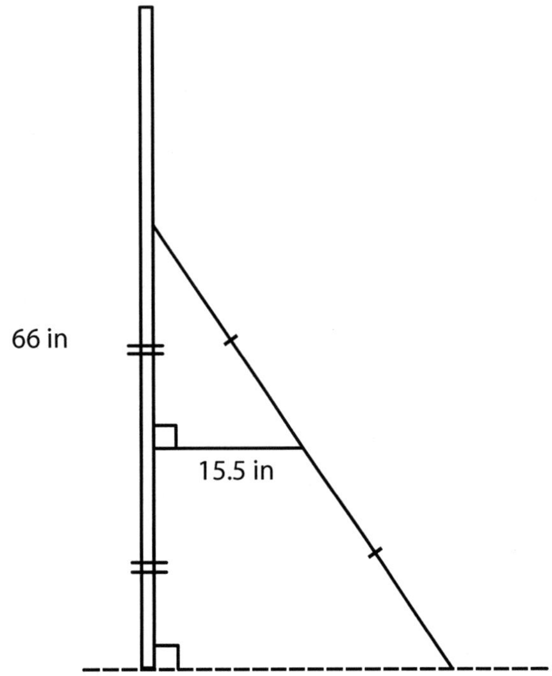

Problem-Based Task: Life-Size Support

Coaching

a. If the cutout measures 66 inches, what is $\frac{2}{3}$ the length of the cutout?

b. The horizontal support acts as a midsegment connecting the cutout to the angled support. How does the length of the midsegment compare to the side it is parallel to?

c. If the horizontal support is 15.5 inches, what is the distance from the bottom of the angled support to the base of the cutout?

d. How can you determine the length of the angled support?

e. What is the length of the angled support rounded to the nearest inch?

Problem-Based Task: Life-Size Support

Coaching Sample Responses

a. If the cutout measures 66 inches, what is $\frac{2}{3}$ the length of the cutout?

To calculate $\frac{2}{3}$ the length of the cutout, multiply $\frac{2}{3}$ by 66.

$$66 \bullet \frac{2}{3} = 44$$

$\frac{2}{3}$ the length of the cutout is 44 inches.

b. The horizontal support acts as a midsegment connecting the cutout to the angled support. How does the length of the midsegment compare to the side it is parallel to?

The midsegment connects the midpoints of two sides of the triangle, is parallel to the third side, and is half as long as the third side.

c. If the horizontal support is 15.5 inches, what is the distance from the bottom of the angled support to the base of the cutout?

The horizontal support is a midsegment of the triangle; therefore, the support is half as long as the distance from the angled support to the base of the cutout.

Multiply 15.5 by 2 to find this distance.

$$15.5 \bullet 2 = 31$$

The distance from the angled support to the base of the cutout is 31 inches.

d. How can you determine the length of the angled support?

The cutout and the angled support create a right triangle, for which the angled support is the hypotenuse.

To find the length of the hypotenuse, use the Pythagorean Theorem.

e. What is the length of the angled support rounded to the nearest inch?

$a^2 + b^2 = c^2$	Pythagorean Theorem
$31^2 + 44^2 = c^2$	Substitute 31 and 44 for the lengths of *a* and *b*.
$961 + 1936 = c^2$	Simplify.
$2897 = c^2$	
$c = \sqrt{2897} \approx 54$ inches	

The length of the angled support rounded to the nearest inch is 54 inches.

Common Core State Standard

G–CO.10

Prove theorems about triangles. *Theorems include: measures of interior angles of a triangle sum to 180°; base angles of isosceles triangles are congruent; the segment joining midpoints of two sides of a triangle is parallel to the third side and half the length; the medians of a triangle meet at a point.*

Problem-Based Task: Sailing Centroid

When sailing, the boat must remain balanced at all times. There are many factors that can affect the stability and balance of a sailboat. One such factor is the determination of the correct center of effort, or the middle of the sail area. This center is found by determining the centroid of the sail. Donzel's sail is a right triangle with a base of 6 feet and a height of 15 feet. The sail is represented on the graph below. Donzel has found what he thinks is the centroid of the sail, marked by point *D* on the graph. Has Donzel correctly identified the centroid?

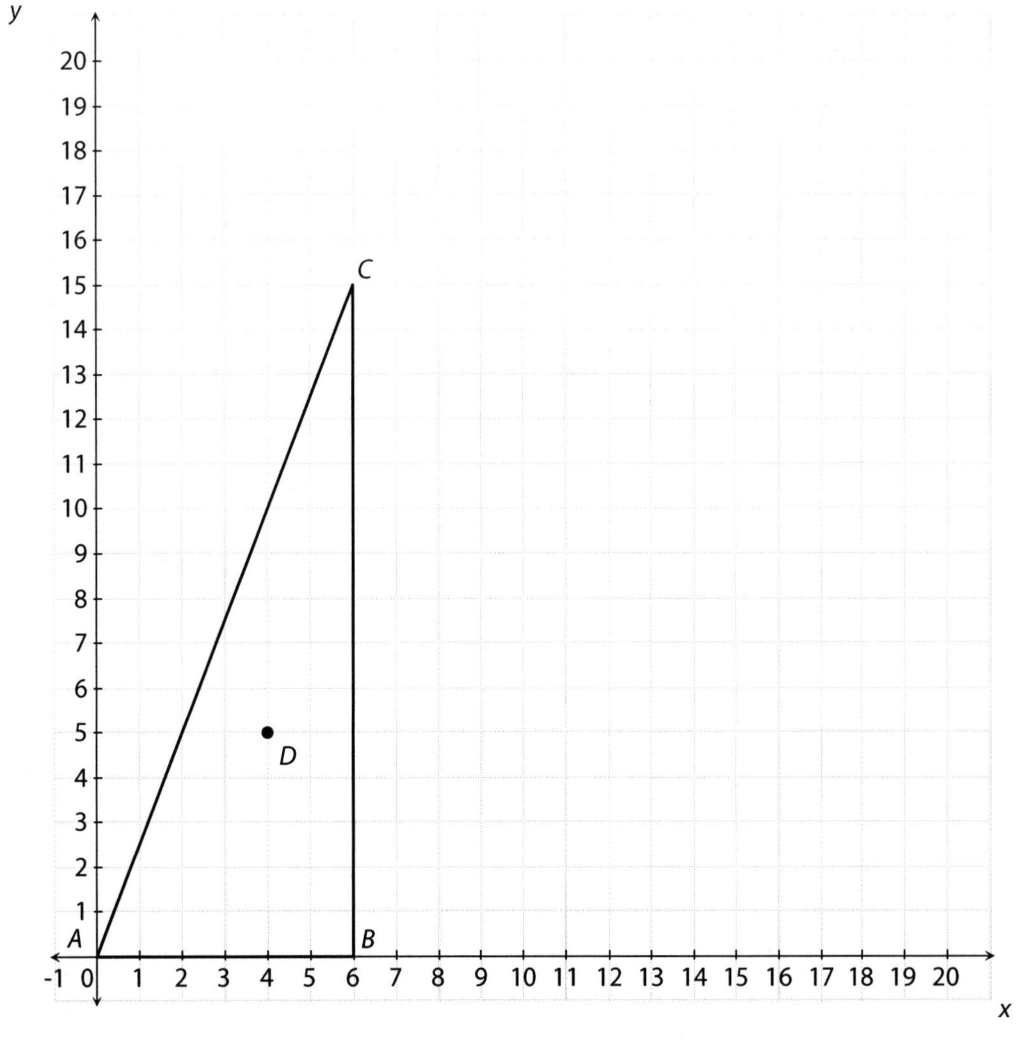

Problem-Based Task: Sailing Centroid

Coaching

a. What is the definition of a centroid?

b. What are the vertices of the triangle representing the sail?

c. What is the midpoint of $\overline{AB}$?

d. What is the midpoint of $\overline{BC}$?

e. What is the midpoint of $\overline{AC}$?

f. What is the equation of the median that passes through $\angle A$?

g. What is the equation of the median that passes through $\angle B$?

h. What is the equation of the median that passes through $\angle C$?

i. Does the point (4, 5) satisfy the equation of each median?

j. Has Donzel correctly identified the centroid?

Problem-Based Task: Sailing Centroid

Coaching Sample Responses

a. What is the definition of a centroid?

The centroid of a triangle is the intersection of the medians of the triangle.

b. What are the vertices of the triangle representing the sail?

The vertices of the triangle representing the sail are A (0, 0), B (6, 0), and C (6, 15).

c. What is the midpoint of $\overline{AB}$?

Use the midpoint formula to find the midpoint of $\overline{AB}$.

$\left(\frac{x_1+x_2}{2},\frac{y_1+y_2}{2}\right)$	Midpoint formula
$\left(\frac{(0)+(6)}{2},\frac{(0)+(0)}{2}\right)$	Substitute (0, 0) and (6, 0) for (x_1, y_1) and (x_2, y_2).
$\left(\frac{6}{2},\frac{0}{2}\right)$	Simplify.
(3, 0)	

The midpoint of $\overline{AB}$ is (3, 0).

d. What is the midpoint of $\overline{BC}$?

Use the midpoint formula to find the midpoint of $\overline{BC}$.

$\left(\frac{x_1+x_2}{2},\frac{y_1+y_2}{2}\right)$	Midpoint formula
$\left(\frac{(6)+(6)}{2},\frac{(0)+(15)}{2}\right)$	Substitute (6, 0) and (6, 15) for (x_1, y_1) and (x_2, y_2).
$\left(\frac{12}{2},\frac{15}{2}\right)$	Simplify.
(6, 7.5)	

The midpoint of $\overline{BC}$ is (6, 7.5).

e. What is the midpoint of $\overline{AC}$?

Use the midpoint formula to find the midpoint of $\overline{AC}$.

$\left(\frac{x_1+x_2}{2}, \frac{y_1+y_2}{2}\right)$	Midpoint formula
$\left(\frac{(0)+(6)}{2}, \frac{(0)+(15)}{2}\right)$	Substitute (0, 0) and (6, 15) for (x_1, y_1) and (x_2, y_2).
$\left(\frac{6}{2}, \frac{15}{2}\right)$	Simplify.
(3, 7.5)	

The midpoint of $\overline{AC}$ is (3, 7.5).

f. What is the equation of the median that passes through *A*?

Find the equation of the line that passes through *A* and the midpoint of $\overline{BC}$.

Use the slope formula to calculate the slope of this equation.

$m=\frac{y_2-y_1}{x_2-x_1}$	Slope formula
$m=\frac{(7.5)-(0)}{(6)-(0)}$	Substitute (0, 0) and (6, 7.5) for (x_1, y_1) and (x_2, y_2).
$m=\frac{7.5}{6}$	Simplify.
$m = 1.25$	

The slope of the line that passes through *A* and the midpoint of $\overline{BC}$ is 1.25.

Find the *y*-intercept of the line that passes through *A* and the midpoint of $\overline{BC}$.

$y-y_1=m(x-x_1)$	Point-slope form of a line
$y-0=1.25(x-0)$	Substitute (0, 0) for (x_1, y_1) and 1.25 for *m*.
$y=1.25x$	Simplify.

The equation of the line that passes through *A* and the midpoint of $\overline{BC}$ is $y = 1.25x$.

g. What is the equation of the median that passes through *B*?

Find the equation of the line that passes through *B* and the midpoint of $\overline{AC}$.

Use the slope formula to calculate the slope of this equation.

$m=\frac{y_2-y_1}{x_2-x_1}$	Slope formula
$m=\frac{(7.5)-(0)}{(3)-(6)}$	Substitute (6, 0) and (3, 7.5) for (x_1, y_1) and (x_2, y_2).
$m=\frac{7.5}{-3}$	Simplify.
$m=-2.5$	

The slope of the line that passes through *B* and the midpoint of $\overline{AC}$ is –2.5.

Find the *y*-intercept of the line that passes through *B* and the midpoint of $\overline{AC}$.

$y-y_1=m(x-x_1)$	Point-slope form of a line
$y-0=-2.5(x-6)$	Substitute (6, 0) for (x_1, y_1) and –2.5 for *m*.
$y-0=-2.5x+15$	Simplify.
$y=-2.5x+15$	

The equation of the line that passes through *B* and the midpoint of $\overline{AC}$ is $y=-2.5x+15$.

h. What is the equation of the median that passes through *C*?

Find the equation of the line that passes through *C* and the midpoint of $\overline{AB}$.

Use the slope formula to calculate the slope of this equation.

$m=\frac{y_2-y_1}{x_2-x_1}$	Slope formula
$m=\frac{(0)-(15)}{(3)-(6)}$	Substitute (6, 15) and (3, 0) for (x_1, y_1) and (x_2, y_2).
$m=\frac{-15}{-3}$	Simplify.
$m=5$	

The slope of the line that passes through *C* and the midpoint of $\overline{AB}$ is 5.

Find the y-intercept of the line that passes through C and the midpoint of $\overline{AB}$.

$y - y_1 = m(x - x_1)$	Point-slope form of a line
$y - 15 = 5(x - 6)$	Substitute (6, 15) for (x_1, y_1) and 5 for m.
$y - 15 = 5x - 30$	Simplify.
$y = 5x - 15$	

The equation of the line that passes through C and the midpoint of $\overline{AB}$ is $y = 5x - 15$.

i. Does the point (4, 5) satisfy the equation of each median?

For (4, 5) to be the intersection of the three medians, the point must satisfy each of the equations: $y = 1.25x$, $y = -2.5x + 15$, and $y = 5x - 15$.

$y = 1.25x$	Equation of the median from A to the midpoint of $\overline{BC}$
$5 = 1.25(4)$	Substitute (4, 5) for (x, y).
$5 = 5$	Simplify.

(4, 5) satisfies the equation of the median from A to the midpoint of $\overline{BC}$.

$y = -2.5x + 15$	Equation of the median from B to the midpoint of $\overline{AC}$
$5 = -2.5(4) + 15$	Substitute (4, 5) for (x, y).
$5 = -10 + 15$	Simplify.
$5 = 5$	

(4, 5) satisfies the equation of the median from B to the midpoint of $\overline{AC}$.

$y = 5x - 15$	Equation of the median from C to the midpoint of $\overline{AB}$
$5 = 5(4) - 15$	Substitute (4, 5) for (x, y).
$5 = 20 - 10$	Simplify.
$5 = 5$	

(4, 5) satisfies the equation of the median from C to the midpoint of $\overline{AB}$.

j. Has Donzel correctly identified the centroid?

Yes; D (4, 5) is the centroid of $\triangle ABC$ with vertices A (0, 0), B (3, 0), and C (6, 15) because (4, 5) satisfies each of the equations of the medians of the triangle.

Common Core State Standard

G–CO.11

Prove theorems about parallelograms. *Theorems include: opposite sides are congruent, opposite angles are congruent, the diagonals of a parallelogram bisect each other, and conversely, rectangles are parallelograms with congruent diagonals.*

Problem-Based Task: Measuring Bridge Cable Tension

A bridge truss is in the shape of a parallelogram as pictured below. To support the truss, steel cables are placed on each of the diagonals of the parallelogram. Engineers are concerned with the tension in one pair of cables, and want to put a strain gauge in the center of the two cables to measure the tension. If the tension is too high, the cables could snap. How will the engineers determine where to place the strain gauges without measuring the length of each cable? Prove that the engineers' decision will work. Use the diagram below for reference in your proof.

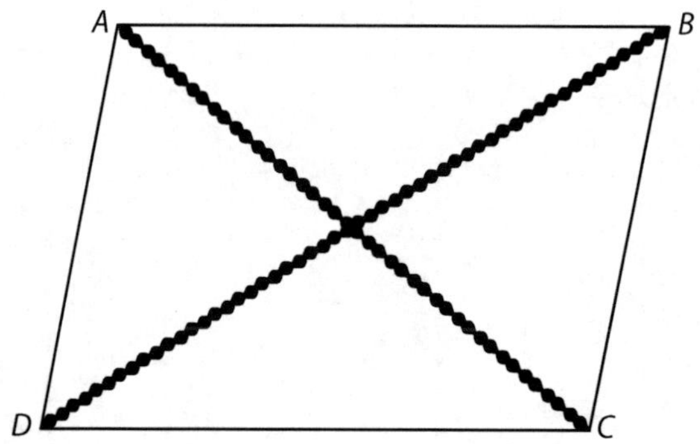

Problem-Based Task: Measuring Bridge Cable Tension

Coaching

a. What is the property of the diagonals of a parallelogram?

b. How can you use this property?

c. Which lines are parallel?

d. What are the transversals?

e. What angles are congruent and how do you know?

f. What sides are congruent and how do you know?

g. What triangles are congruent and how do you know?

h. What segments are congruent that you have not already identified? How do you know?

i. Summarize your findings and write your proof.

Problem-Based Task: Measuring Bridge Cable Tension

Coaching Sample Responses

a. What is the property of the diagonals of a parallelogram?

The diagonals of a parallelogram bisect each other.

b. How can you use this property?

Where the cables intersect is the midpoint of each cable and, therefore, the center of the cable. That is where the strain gauge should be located.

c. Which lines are parallel?

Since the truss *ABCD* is a parallelogram, opposite sides are parallel. This means that the top is parallel to the bottom and the left side is parallel to the right side; or, stated mathematically, $\overline{AB} \| \overline{DC}$ and $\overline{AD} \| \overline{BC}$.

d. What are the transversals?

Each cable acts as a transversal to both sets of parallel lines. This means that $\overline{AC}$ is a transversal of $\overline{AB}$ and $\overline{DC}$, as well as a transversal of $\overline{AD}$ and $\overline{BC}$. This also means that $\overline{DB}$ is a transversal of $\overline{AB}$ and $\overline{DC}$, as well as a transversal of $\overline{AD}$ and $\overline{BC}$.

e. What angles are congruent and how do you know?

Because alternate interior angles of a set of parallel lines intersected by a transversal are congruent, $\angle BAC \cong \angle DCA$, $\angle BDC \cong \angle DBA$, $\angle DAC \cong \angle ACB$, and $\angle ADB \cong \angle DBC$.

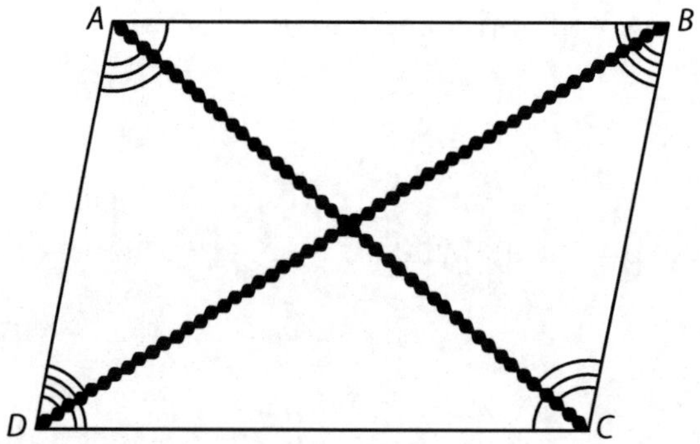

f. What sides are congruent and how do you know?

Opposite sides of a parallelogram are congruent. Therefore, $\overline{AB} \cong \overline{DC}$ and $\overline{AD} \cong \overline{BC}$.

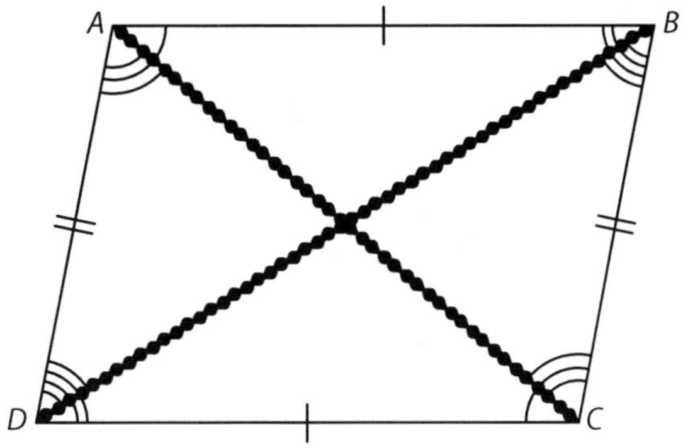

g. What triangles are congruent and how do you know?

Label the intersection point of the diagonals with the letter *P*.

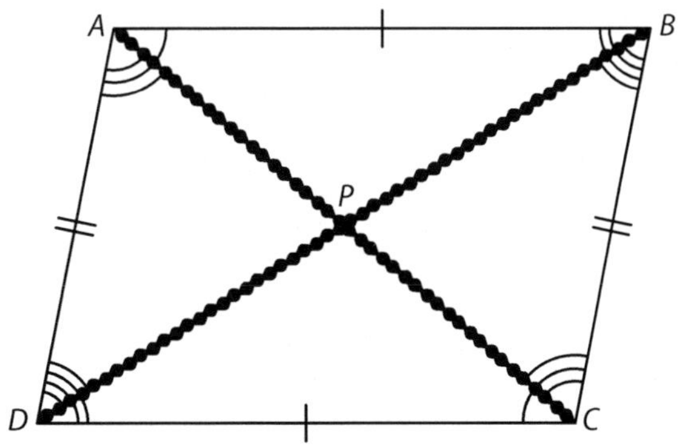

By SAS, $\triangle ABP \cong \triangle CDP$ and $\triangle ADP \cong \triangle CBP$.

h. What segments are congruent that you have not already identified? How do you know?

Because of CPCTC, $\overline{AP} \cong \overline{CP}$ and $\overline{DP} \cong \overline{BP}$. By the definition of bisection, $\overline{AC}$ bisects $\overline{DB}$ and $\overline{DB}$ bisects $\overline{AC}$.

i. Summarize your findings and write your proof.

Given that *ABCD* is a parallelogram, opposite sides are parallel, so $\overline{AB} \parallel \overline{DC}$ and $\overline{AD} \parallel \overline{BC}$. Since alternate interior angles of a set of parallel lines intersected by a transversal are congruent, $\angle BAC \cong \angle DCA$, $\angle BDC \cong \angle DBA$, $\angle DAC \cong \angle ACB$, and $\angle ADB \cong \angle DBC$. Opposite sides of a parallelogram are congruent. Therefore, $\overline{AB} \cong \overline{DC}$ and $\overline{AD} \cong \overline{BC}$. By SAS, $\triangle ABP \cong \triangle CDP$ and $\triangle ADP \cong \triangle CBP$. Since corresponding parts of congruent triangles are congruent, $\overline{AP} \cong \overline{CP}$ and $\overline{DP} \cong \overline{BP}$. By the definition of bisection, $\overline{AC}$ bisects $\overline{DB}$ and $\overline{DB}$ bisects $\overline{AC}$.

G–CO.11 • Geometry

Proving Properties of Special Quadrilaterals

Common Core State Standard

G–CO.11

Prove theorems about parallelograms. *Theorems include: opposite sides are congruent, opposite angles are congruent, the diagonals of a parallelogram bisect each other, and conversely, rectangles are parallelograms with congruent diagonals.*

Problem-Based Task: Where Are the Catchers?

Aric, Carl, Bree, and Daisy play Little League baseball. At catching practice, the coach positions each player according to his or her skill level. The four players' positions form a rhombus. Aric, Bree, and Carl can be represented on a coordinate plane by points A (0, 0), B (–5, 0), and C (–1, 3). Find Daisy's location, point D.

Elian and Faith are the best catchers on the team, so the coach has them practice separately. The coach places Elian in a spot represented by point E (7, –6) on a coordinate plane. The coach places Faith so that she's as far away from Elian as Bree is from Daisy. If Bree, Daisy, Elian, and Faith's positions form a rectangle, find point F, Faith's position. Graph the positions of all the players.

Problem-Based Task: Where Are the Catchers?

Coaching

a. What does the graph of the given points look like?

b. What do you know about the diagonals of a rhombus?

c. What is the slope of $\overline{CA}$?

d. What is the slope of $\overline{BD}$ and why?

e. What is the equation of the line that encompasses $\overline{BD}$?

f. What other line intersects $\overline{BD}$ where we know the slope and one point?

g. What is the slope of $\overline{AD}$, and what given point is on that line?

h. What is the equation of $\overline{AD}$?

i. How can you use these two equations to find point *D*?

j. What is point *D*?

k. What point is still unknown?

l. What do we know about the slopes of the sides of a rectangle?

m. What is the slope of $\overline{BD}$?

n. What is the slope of $\overline{EF}$?

o. What is the equation of $\overline{EF}$?

p. What other line intersects $\overline{EF}$ where we know the slope and one point?

q. What is the equation of this line?

r. How can you use the two equations to find point *F*?

s. What is point *F*?

t. What is the graph of the final positions of all the catchers?

Problem-Based Task: Where Are the Catchers?

Coaching Sample Responses

a. What does the graph of the given points look like?

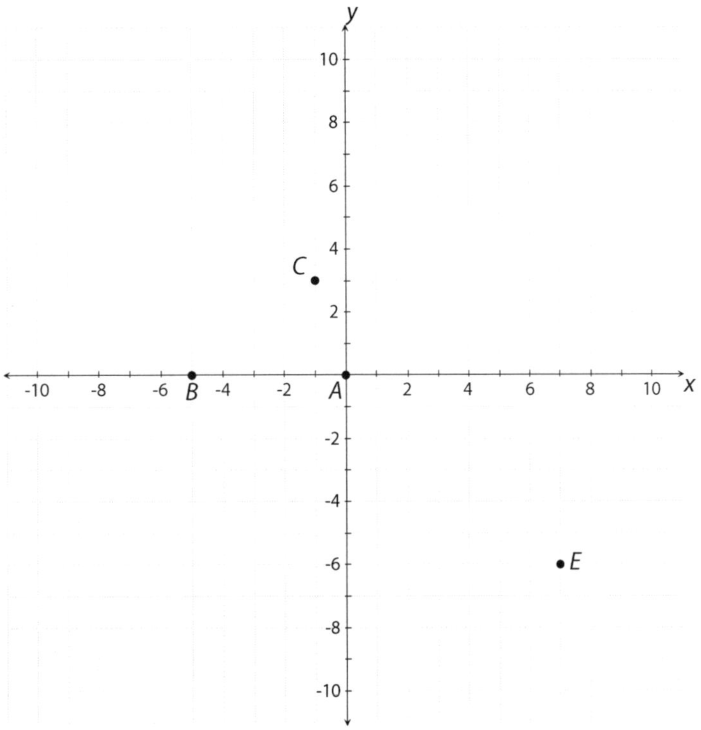

b. What do you know about the diagonals of a rhombus?

They intersect at right angles, so the lines are perpendicular, and the slopes are opposite reciprocals.

c. What is the slope of $\overline{CA}$?

$$m_{\overline{CA}} = \frac{\Delta y}{\Delta x} = \frac{0-3}{0-(-1)} = -\frac{3}{1} = -3$$

d. What is the slope of $\overline{BD}$ and why?

$m_{\overline{BD}} = \frac{1}{3}$ because it is the opposite reciprocal of $\overline{CA}$, since $\overline{CA} \perp \overline{BD}$.

e. What is the equation of the line that encompasses $\overline{BD}$?

We know that point B (–5, 0) is on the line and the slope is $\frac{1}{3}$. Use the slope-intercept form of the equation to find the equation of the line.

$$y=mx+b$$
$$0=\frac{1}{3}(-5)+b$$
$$0=-\frac{5}{3}+b$$
$$b=\frac{5}{3}$$

$$y=mx+b$$
$$y=\frac{1}{3}x+\frac{5}{3}$$

f. What other line intersects $\overline{BD}$ where we know the slope and one point?

There are two other lines: $\overline{CD}$ and $\overline{AD}$.

g. What is the slope of $\overline{AD}$, and what given point is on that line?

$\overline{AD}$ is parallel to $\overline{BC}$ because opposite pairs of sides in a parallelogram are parallel and a rhombus is a parallelogram. When lines are parallel, their slopes are equal.

$$m_{\overline{BC}}=\frac{\Delta y}{\Delta x}=\frac{3-0}{-1-(-5)}=\frac{3}{4}$$
$$m_{\overline{AD}}=\frac{3}{4}$$

Point A is on the line and is given, A (0, 0).

h. What is the equation of $\overline{AD}$?

Insert the slope of $\overline{AD}$ and the point A (0, 0) into the equation $y = mx + b$.

$$y=mx+b$$
$$0=\frac{3}{4}(0)+b$$
$$0=0+b$$
$$b=0$$

$$y=mx+b$$
$$y=\frac{3}{4}x+0$$
$$y=\frac{3}{4}x$$

i. How can you use these two equations to find point D?

Solve the two equations as a system of equations. The solution will be the intersection point of the two lines, point D.

j. What is point D?

Set the two equations equal to each other since they are both equal to y.

$$\overline{BD}: y=\frac{1}{3}x+\frac{5}{3}$$

$$\overline{AD}: y=\frac{3}{4}x$$

Find x.

$$\frac{1}{3}x+\frac{5}{3}=\frac{3}{4}x$$

$$\frac{1}{3}x-\frac{3}{4}x=-\frac{5}{3}$$

$$\frac{4}{12}x-\frac{9}{12}x=-\frac{5}{3}$$

$$-\frac{5}{12}x=-\frac{5}{3}$$

$$x=-\frac{5}{3}\bullet-\frac{12}{5}$$

$$x=\frac{60}{15}$$

$$x=4$$

Use x to find y.

$$y=\frac{3}{4}x$$

$$y=\frac{3}{4}(4)$$

$$y=3$$

Point D is at (4, 3).

k. What point is still unknown?

Point F is the second unknown point.

l. What do we know about the slopes of the sides of a rectangle?

The opposite sides are parallel and consecutive sides are perpendicular. The opposite sides have the same slope and the consecutive sides have slopes that are opposite reciprocals.

m. What is the slope of $\overline{BD}$?

From part c, we know $m_{\overline{BD}} = \frac{1}{3}$.

n. What is the slope of $\overline{EF}$?

$\overline{EF}$ is parallel to $\overline{BD}$, so the slopes are equal.

$$m_{\overline{EF}} = \frac{1}{3}$$

o. What is the equation of $\overline{EF}$?

We can determine the equation of the line if we know the slope and a point on the line. We know the slope, $\frac{1}{3}$, and a point on the line, E (7, –6). Substitute these into $y = mx + b$.

$$y = mx + b$$
$$-6 = \frac{1}{3}(7) + b$$
$$-6 = \frac{7}{3} + b$$
$$-6 - \frac{7}{3} = b$$
$$-\frac{18}{3} - \frac{7}{3} = b$$
$$b = -\frac{25}{3}$$

$$y = mx + b$$
$$y = \frac{1}{3}x - \frac{25}{3}$$

p. What other line intersects $\overline{EF}$ where we know the slope and one point?

$\overline{BF}$ intersects $\overline{EF}$.

q. What is the equation of the line?

We know the slope of $\overline{BF}$ is perpendicular to $\overline{EF}$. Therefore, the slopes are opposite reciprocals. The slope of $\overline{EF}$ is $\frac{1}{3}$. This means the slope of $\overline{BF}$ is –3. We know point *B* (–5, 0). Substitute this information into the slope-intercept form of the equation.

$$y=mx+b$$
$$0=-3(-5)+b$$
$$0=15+b$$
$$b=-15$$

$$y=mx+b$$
$$y=-3x-15$$

r. How can you use the two equations to find point *F*?

Solve the two equations as a system. The solution will be the intersection point, *F*.

s. What is point *F*?

Set the equations equal to each other since they are both equal to *y*.

$$\overline{EF}: y=\frac{1}{3}x-\frac{25}{3}; \overline{BF}: y=-3x-15$$

Find *x*.

$$\frac{1}{3}x-\frac{25}{3}=-3x-15$$
$$\frac{1}{3}x+3x=-15+\frac{25}{3}$$
$$\frac{1}{3}x+\frac{9}{3}x=-\frac{45}{3}+\frac{25}{3}$$
$$\frac{10}{3}x=-\frac{20}{3}$$
$$x=-\frac{20}{3}\bullet\frac{3}{10}$$
$$x=-\frac{60}{30}$$
$$x=-2$$

Find *y*.

$$y=-3x-15$$
$$y=-3(-2)-15$$
$$y=6-15$$
$$y=-9$$

Point *F* is at (–2, –9).

t. What is the graph of the final positions of all the catchers?

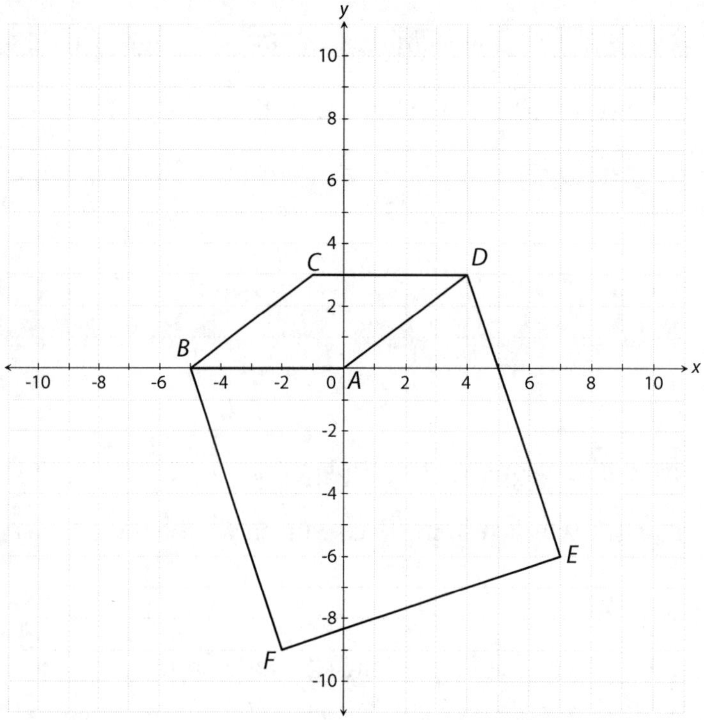

G–SRT.1a • Geometry

Investigating Properties of Parallelism and the Center

Common Core State Standard

G–SRT.1

Verify experimentally the properties of dilations given by a center and a scale factor:

a. A dilation takes a line not passing through the center of the dilation to a parallel line, and leaves a line passing through the center unchanged.

Problem-Based Task: Prettying Up the Pentagon

The Pentagon, diagrammed below, is one of the world's largest office buildings. The outside walls are each 921 feet long and are a dilation of the inner walls through the center of the courtyard. The courtyard is the area inside the inner wall. Since the courtyard is surrounded by the inner walls, each side of the courtyard is the same length as the inner walls. The dashed lines represent a walkway that borders a garden. The walkway is a dilation of the inner wall of the office building.

A team of landscapers has been hired to update the courtyard. The landscapers need to know the perimeter of the walkway in order to install some temporary fencing while the courtyard is redone. What is the perimeter of the walkway if the dilation from the inner wall to the walkway has a scale factor of 0.25? What relationship does the scale factor have to the perimeters of the figures?

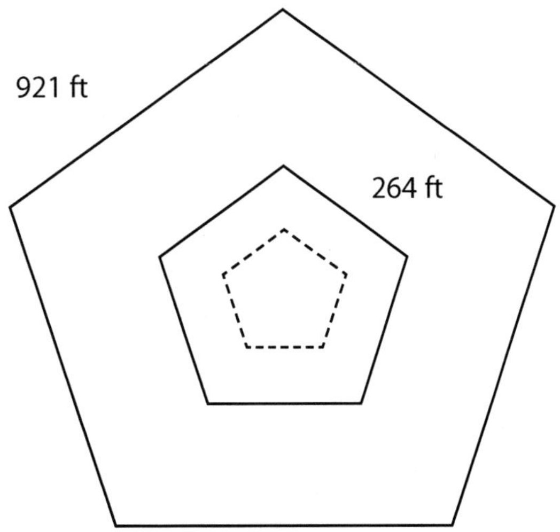

Problem-Based Task: Prettying Up the Pentagon

Coaching

a. What is the length of one side of the inner wall (the preimage)?

b. What is the scale factor used to dilate the inner wall to determine the walkway?

c. How do you calculate the length of each side of the walkway in the pentagonal figure in the courtyard?

d. What is the length of each side of the walkway?

e. How many sides are there in a pentagon?

f. What is the perimeter of the walkway?

g. What is the perimeter of the inner wall?

h. What is the scale factor of the perimeter of the inner wall to the perimeter of the walkway?

i. How do the scale factors of the lengths of each side of the inner wall and the walkway compare to the perimeters of the inner wall and the walkway?

j. What is the scale factor of the inner wall to the outer wall?

k. What is the perimeter of the outer wall?

l. What is the scale factor of the perimeter of the inner wall to the perimeter of the outer wall?

m. How do the scale factors of the individual side lengths of the inner and outer walls compare to the perimeters of the inner and outer walls?

n. What can you conclude about the scale factors of perimeters of dilated figures?

Problem-Based Task: Prettying Up the Pentagon

Coaching Sample Responses

a. What is the length of one side of the inner wall (the preimage)?

264 feet

b. What is the scale factor used to dilate the inner wall to determine the walkway?

0.25

c. How do you calculate the length of each side of the walkway in the pentagonal figure in the courtyard?

Multiply the preimage side by the scale factor: 264(0.25).

d. What is the length of each side of the walkway?

$264(0.25) = 66$

The length of each side of the walkway in the pentagonal figure is 66 feet.

e. How many sides are there in a pentagon?

A pentagon has 5 sides.

f. What is the perimeter of the walkway?

$5(66) = 330$

The perimeter of the walkway is 330 feet.

g. What is the perimeter of the inner wall?

$5(264) = 1320$

The perimeter of the inner wall is 1,320 feet.

h. What is the scale factor of the perimeter of the inner wall to the perimeter of the walkway?

$$\frac{\text{perimeter of walkway}}{\text{perimeter of inner wall}} = \frac{330}{1320} = 0.25$$

i. How do the scale factors of the lengths of each side of the inner wall and the walkway compare to the perimeters of the inner wall and the walkway?

The scale factor of each wall is the same as the scale factor of the perimeters.

j. What is the scale factor of the inner wall to the outer wall?

$$\frac{\text{length of inner wall}}{\text{length of outer wall}}=\frac{264}{921}\approx 0.29$$

k. What is the perimeter of the outer wall?

$5(921) = 4605$

The perimeter of the outer wall is 4,605 feet.

l. What is the scale factor of the perimeter of the inner wall to the perimeter of the outer wall?

$$\frac{\text{perimeter of inner wall}}{\text{perimeter of outer wall}}=\frac{1320}{4605}\approx 0.29$$

m. How do the scale factors of the individual side lengths of the inner and outer walls compare to the perimeters of the inner and outer walls?

The scale factors are the same.

n. What can you conclude about the scale factors of perimeters of dilated figures?

If a figure is dilated, the perimeter of the preimage to the image has the same scale factor as the dilation.

Common Core State Standard

G–SRT.1

Verify experimentally the properties of dilations given by a center and a scale factor:

b. The dilation of a line segment is longer or shorter in the ratio given by the scale factor.

Problem-Based Task: The Bigger Picture

A photographer wants to enlarge a 5×7 picture to an 8×10. However, she wants to preserve the image as it appears in the 5×7 without distorting the picture. Distortions happen when the width and height of the photo are not enlarged at the same scale. How can the photographer dilate a 5×7 picture to an 8×10 picture without distorting the picture? Describe a process for enlarging the picture so that the image is a dilation of the preimage. Give the coordinates for the image vertices. The preimage is pictured below with the center C (0, 0).

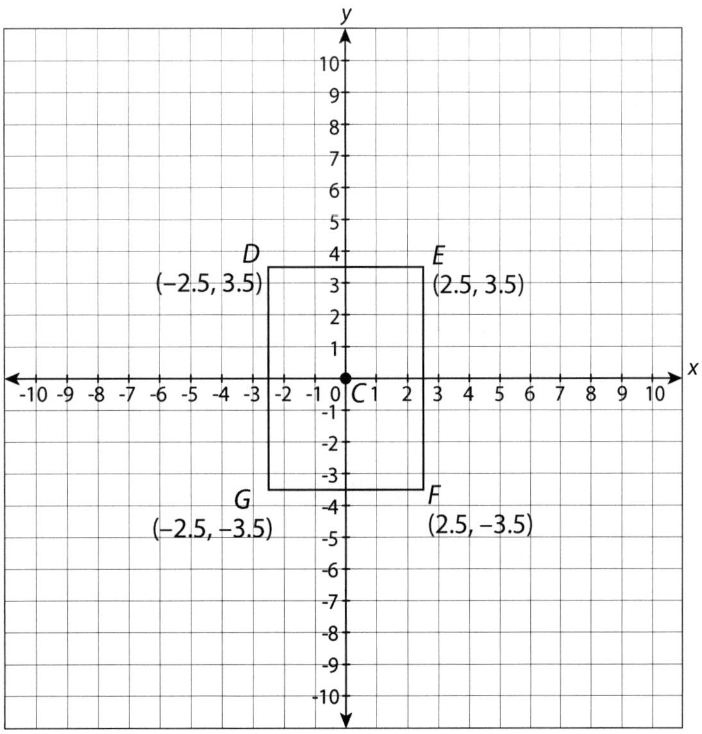

Problem-Based Task: The Bigger Picture

Coaching

a. What is the scale factor of the width from the preimage to the image?

b. What is the scale factor of the height from the preimage to the image?

c. How do these scale factors compare?

d. Which scale factor can you use consistently with the width and the height so that the picture will not be distorted?

e. How can you modify the preimage so that you can use the same scale factor and arrive at an 8×10 picture?

f. How can you determine numerically how to modify the picture?

g. What are the coordinates of the modified preimage?

h. What are the coordinates of the image after applying the scale factor?

i. Summarize your procedure and thinking process in dilating the original photograph.

Problem-Based Task: The Bigger Picture

Coaching Sample Responses

a. What is the scale factor of the width from the preimage to the image?

$$\frac{\text{width of image}}{\text{width of preimage}}=\frac{8}{5}=1.6$$

b. What is the scale factor of the height from the preimage to the image?

$$\frac{\text{height of image}}{\text{height of preimage}}=\frac{10}{7}\approx 1.429$$

c. How do these scale factors compare?

The scale factor of the height is smaller than the scale factor for the width.

d. Which scale factor can you use consistently with the width and the height so that the picture will not be distorted?

Examine what happens when you apply the scale factor backward. In other words, find the reciprocal of each scale factor.

The scale factor for the width = 8/5. The reciprocal is 5/8 or 0.625.

The scale factor for the height = 10/7. The reciprocal is 7/10 or 0.7.

Use the smaller reciprocal scale factor since you cannot generate more of an original picture. This means when you go from the smaller preimage to the image, you will use the reciprocal of 5/8, which is 8/5, the scale factor of the width.

e. How can you modify the preimage so that you can use the same scale factor and arrive at an 8×10 picture?

Crop or trim the picture width so that the preimage width becomes smaller.

f. How can you determine numerically how to modify the picture?

To determine by how much to trim the picture, set up a proportion with the unknown being the height of the original picture. Since we are using one scale factor for both dimensions, the second ratio is the ratio of the width of the image to the width of the preimage.

$$\frac{10}{x}=\frac{8}{5}$$

$$50=8x$$

$$x=6.25$$

The original height is 7 inches. The picture needs to be trimmed by 7 – 6.25 inches, or 0.75 inches. The image could be trimmed 0.75 inches at the bottom or the top or by a combination to avoid losing any of the important aspects of the picture.

g. What are the coordinates of the modified preimage?

Assuming equal trimming of the top and bottom of the picture, the amount to trim at each end is 0.75/2 = 0.375 inches. For negative coordinates, you must add 0.375 to show the trimming.

D: 3.5 – 0.375 = 3.125

E: 3.5 – 0.375 = 3.125

F: –3.5 + 0.375 = –3.125

G: –3.5 + 0.375 = –3.125

The coordinates of the modified preimage are as follows:

$$D(-2.5, 3.5) \to D_m(-2.5, 3.125)$$
$$E(2.5, 3.5) \to E_m(2.5, 3.125)$$
$$F(2.5, -3.5) \to F_m(2.5, -3.125)$$
$$G(-2.5, -3.5) \to G_m(-2.5, -3.125)$$

h. What are the coordinates of the image after applying the scale factor?

Apply the scale factor of 8/5 or 1.6 to each coordinate in the modified preimage.

$$D_m(-2.5, 3.125) \to D'_m(-4, 5)$$

$$E_m(2.5, 3.125) \to E'_m(4, 5)$$

$$F_m(2.5, -3.125) \to F'_m(4, -5)$$

$$G_m(-2.5, -3.125) \to G'_m(-4, -5)$$

i. Summarize your procedure and thinking process in dilating the original photograph.

First, determine the scale factor to use by calculating the scale factor for each dimension of width and height. Use the scale factor that will allow you to trim the photograph rather than adding onto it. This means you use the scale factor with the smaller reciprocal because you will be going backward to determine how much to cut from the picture. The scale factor to use is the scale factor of the width, which is 8/5 or 1.6. The unknown is how long the image should be before dilation. You know that you want to end up with 10 inches in height. Apply the scale factor to 10 inches and the result is 6.25 inches. The picture height is 7 inches.

To modify the height of the picture, trim 0.75 inches from the picture. This can be done equally from the top and bottom (0.375 inches from the top and 0.375 inches from the bottom), from the top only, from the bottom only, or some combination of the top and bottom that is not equal. For simplicity, trim 0.375 inches from the top and the bottom. On the graph, this means moving each of the y-coordinates 0.375 units toward 0 parallel to the y-axis. The x-coordinates will remain the same because the width is not changing. Now, you apply the scale factor of 8/5 to each coordinate in all the vertices. The resulting image is a dilation of the trimmed picture.

G–SRT.2 • Geometry
Defining Similarity

Common Core State Standard

G–SRT.2

Given two figures, use the definition of similarity in terms of similarity transformations to decide if they are similar; explain using similarity transformations the meaning of similarity for triangles as the equality of all corresponding pairs of angles and the proportionality of all corresponding pairs of sides.

Problem-Based Task: Video Game Transformations

The creators of a new video game are in the early design stages and are using the right triangle *ABC* on a coordinate plane to represent the movement of the character in the actual game setting. The coordinates of the points are A (–6, –9), B (–3, –9), and C (–3, –6). After a series of similarity transformations, the locations of the end points of the hypotenuse of the new image are A' (6, –4) and C' (4, –2). Is it possible to determine the location of point B'?

Problem-Based Task: Video Game Transformations

Coaching

a. What is the length of hypotenuse $\overline{AC}$?

b. What is the length of $\overline{A'C'}$?

c. Is the length of $\overline{A'C'}$ shorter or longer than the length of $\overline{AC}$?

d. By what scale factor was the length of $\overline{AC}$ changed?

e. If a series of similarity transformations were carried out, what would you know about the lengths of the remaining sides of the image compared to the lengths of the preimage?

f. What is the length of $\overline{AB}$?

g. What is the length of $\overline{A'B'}$?

h. What is the length of $\overline{BC}$?

i. What is the length of $\overline{B'C'}$?

j. What are the possible locations of point B'?

Problem-Based Task: Video Game Transformations

Coaching Sample Responses

a. What is the length of hypotenuse $\overline{AC}$?

Use the distance formula to calculate the length of $\overline{AC}$.

$d=\sqrt{(x_2-x_1)^2+(y_2-y_1)^2}$	Distance formula
$d=\sqrt{[(-3)-(-6)]^2+[(-6)-(-9)]^2}$	Substitute (–6, –9) and (–3, –6) for (x_1, y_1) and (x_2, y_2).
$d=\sqrt{(3)^2+(3)^2}$	Simplify.
$d=\sqrt{18}=\sqrt{9\bullet 2}=3\sqrt{2}$	

The length of $\overline{AC}$ is $3\sqrt{2}$ units.

b. What is the length of $\overline{A'C'}$?

Use the distance formula to calculate the length of $\overline{A'C'}$.

$d=\sqrt{(x_2-x_1)^2+(y_2-y_1)^2}$	Distance formula
$d=\sqrt{[(4)-(6)]^2+[(-2)-(-4)]^2}$	Substitute (6, –4) and (4, –2) for (x_1, y_1) and (x_2, y_2).
$d=\sqrt{(-2)^2+(2)^2}$	Simplify.
$d=\sqrt{8}=\sqrt{4\bullet 2}=2\sqrt{2}$	

The length of $\overline{A'C'}$ is $2\sqrt{2}$ units.

c. Is the length of $\overline{A'C'}$ shorter or longer than the length of $\overline{AC}$?

The length of $\overline{AC}$ is $3\sqrt{2}$.

The length of $\overline{A'C'}$ is $2\sqrt{2}$ units.

The length of $\overline{A'C'}$ is shorter than the length of $\overline{AC}$.

d. By what scale factor was the length of $\overline{AC}$ changed?

Divide the length of $\overline{A'C'}$ by the length of $\overline{AC}$.

$$\frac{A'C'}{AC}=\frac{2\sqrt{2}}{3\sqrt{2}}=\frac{2}{3}$$

$\overline{A'C'}$ is $\frac{2}{3}$ the length of $\overline{AC}$.

e. If a series of similarity transformations were carried out, what would you know about the lengths of the remaining sides of the image compared to the lengths of the preimage?

The lengths of the sides of the image are all proportional to the preimage.

Each side of $\triangle A'B'C'$ is $\frac{2}{3}$ the length of the sides of $\triangle ABC$.

f. What is the length of $\overline{AB}$?

Use the distance formula to calculate the length of $\overline{AB}$.

$d=\sqrt{(x_2-x_1)^2+(y_2-y_1)^2}$	Distance formula
$d=\sqrt{[(-3)-(-6)]^2+[(-9)-(-9)]^2}$	Substitute (–6, –9) and (–3, –9) for (x_1, y_1) and (x_2, y_2).
$d=\sqrt{(3)^2+(0)^2}$	Simplify.
$d=\sqrt{9}=3$	

The length of $\overline{AB}$ is 3 units.

g. What is the length of $\overline{A'B'}$?

Use the scale factor of $\frac{2}{3}$ to find the length of $\overline{A'B'}$.

$$\frac{2}{3}\bullet 3=2$$

The length of $\overline{A'B'}$ is 2 units.

h. What is the length of $\overline{BC}$?

Use the distance formula to calculate the length of $\overline{BC}$.

$d=\sqrt{(x_2-x_1)^2+(y_2-y_1)^2}$	Distance formula
$d=\sqrt{[(-3)-(-3)]^2+[(-6)-(-9)]^2}$	Substitute (–3, –9) and (–3, –6) for (x_1, y_1) and (x_2, y_2).
$d=\sqrt{(0)^2+(3)^2}$	Simplify.
$d=\sqrt{9}=3$	

The length of $\overline{BC}$ is 3 units.

i. What is the length of $\overline{B'C'}$?

Use the scale factor of $\frac{2}{3}$ to find the length of $\overline{B'C'}$.

$$\frac{2}{3}\bullet 3=2$$

The length of $\overline{B'C'}$ is 2 units.

j. What are the possible locations of point B'?

There are two possible locations for point B'.

One location for point B' is 2 units to the right of C' and 2 units up from A'.

This location is at (4, –4).

The second location is 2 units down from C' and 2 units to the left of A'.

This location is at (6, –2).

NAME:

G–SRT.3 • Geometry

Applying Similarity Using the Angle-Angle (AA) Criterion

Common Core State Standard

G–SRT.3

Use the properties of similarity transformations to establish the AA criterion for two triangles to be similar.

Problem-Based Task: True Trusses

A mono truss is a type of building support structure that is in the shape of a right triangle. Contractors often use mono trusses when building roofs for small structures such as garages and sheds. The vertical pieces of this truss form $90°$ angles with the horizontal pieces in order to maximize the stability. Observe the diagram of a mono truss below. Is $\triangle ABC$ similar to $\triangle ADE$? Explain your reasoning. Is it possible to determine the length of $\overline{DE}$ from the given information? If so, calculate the length.

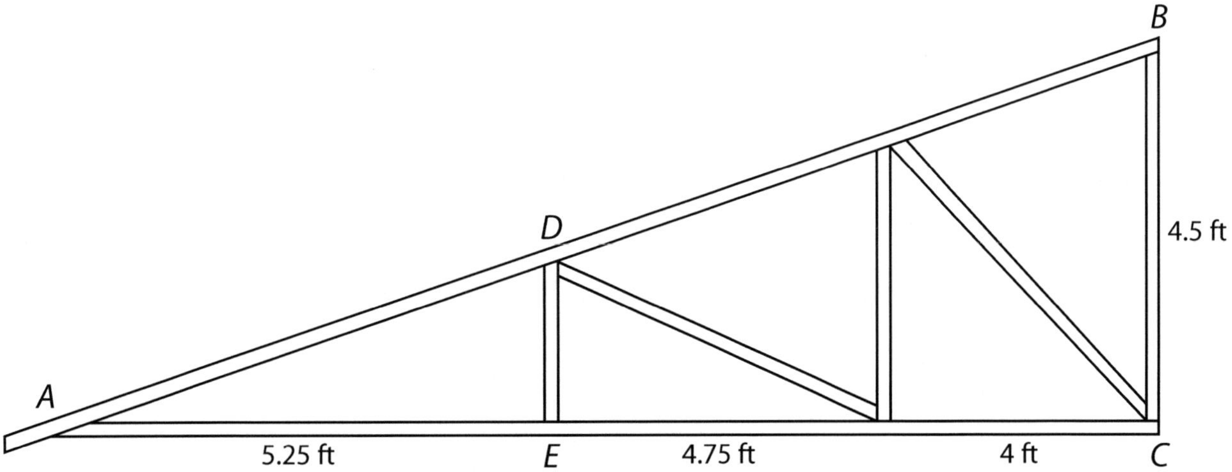

Problem-Based Task: True Trusses

Coaching

a. Which angles of $\triangle ABC$ are congruent to angles of $\triangle ADE$? Name them, and explain why they are congruent.

b. Is there enough information to determine if $\triangle ABC$ is similar to $\triangle ADE$?

c. If two triangles are similar, what must be true about the lengths of their corresponding sides?

d. Identify the corresponding sides of $\triangle ABC$ and $\triangle ADE$.

e. What is the ratio between the length of $\overline{AC}$ and the length of $\overline{AE}$?

f. What is the length of $\overline{DE}$?

Problem-Based Task: True Trusses

Coaching Sample Responses

a. Which angles of $\triangle ABC$ are congruent to angles of $\triangle ADE$? Name them, and explain why they are congruent.

$\angle BCA$ of $\triangle ABC$ is congruent to $\angle DEA$ of $\triangle ADE$ because they are both right angles.

$\angle BAC$ of $\triangle ABC$ is congruent to $\angle DAE$ of $\triangle ADE$ because they are composed of the same angle, $\angle A$.

b. Is there enough information to determine if $\triangle ABC$ is similar to $\triangle ADE$?

Yes; according to the Angle-Angle (AA) Similarity Statement, if two angles of one triangle are congruent to two angles of another triangle, then the triangles are similar.

c. If two triangles are similar, what must be true about the lengths of their corresponding sides?

Similar triangles have corresponding sides that are proportional.

d. Identify the corresponding sides of $\triangle ABC$ and $\triangle ADE$.

$\overline{AB} \rightarrow \overline{AD}$

$\overline{AC} \rightarrow \overline{AE}$

$\overline{BC} \rightarrow \overline{DE}$

e. What is the ratio between the length of $\overline{AC}$ and the length of $\overline{AE}$?

$$\frac{AC}{AE} = \frac{(5.25 + 4.75 + 4)}{5.25} = \frac{14}{5.25} = \frac{8}{3}$$

f. What is the length of $\overline{DE}$?

Create and solve a proportion using the ratio found in part e.

$$\frac{BC}{DE} = \frac{8}{3} = \frac{4.5}{x}$$

$x = 1.6875$

The length of $\overline{DE}$ is 1.6875 feet.

G–SRT.4 • Geometry

Proving Triangle Similarity Using Side-Angle-Side (SAS) and Side-Side-Side (SSS) Similarity

Common Core State Standard

G–SRT.4

Prove theorems about triangles. *Theorems include: a line parallel to one side of a triangle divides the other two sides proportionally, and conversely; the Pythagorean Theorem proved using triangle similarity.*

Problem-Based Task: Down, Down, Down

Gutters are designed to channel rainwater away from roofs, preventing water damage and leaks. Downspouts lead down from the gutters to the ground. The amount of downspout needed depends on the height of the house. If the downspout for the gutters is too short, the water can cause flooding. Blair is using a mirror to determine the height of his house so he can know how long of a downspout he needs to buy. He has placed the mirror on the ground and is standing far enough away from it that he can see the gutters of his house in the mirror. Blair, who is 5 foot 10 inches tall, measured his distance from the mirror as 105 inches. The distance from the mirror to the foundation of the house is 450 inches. Downspouts are often sold in 5-foot sections. How many sections of downspout does Blair need to purchase?

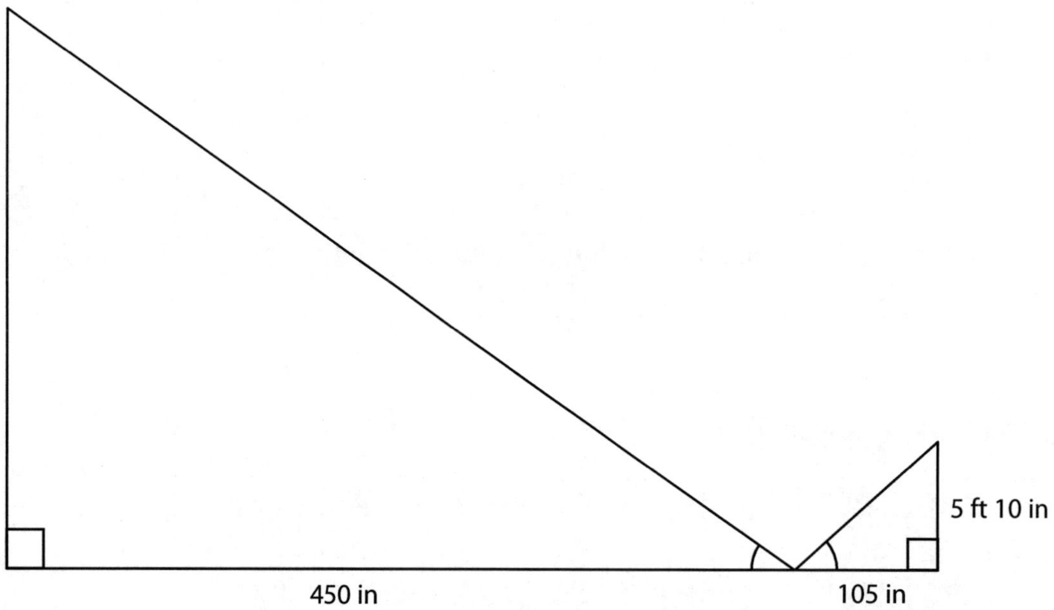

Problem-Based Task: Down, Down, Down

Coaching

a. How tall is Blair in inches?

b. Are the triangles in the diagram similar? Explain.

c. What are the characteristics of similar triangles?

d. Is it possible to determine the height of the house from the foundation to the gutters using the given information? Explain your answer.

e. What is the height of the house from the foundation to the gutters?

f. How many 5-foot sections of downspout are needed for Blair's house?

Problem-Based Task: Down, Down, Down

Coaching Sample Responses

a. How tall is Blair in inches?

There are 12 inches in a foot. Blair is 5 feet 10 inches tall.

To determine his height in inches, multiply the number of feet tall he is by 12, and then add the number of inches.

height in inches = $5(12) + 10 = 60 + 10 = 70$

Blair is 70 inches tall.

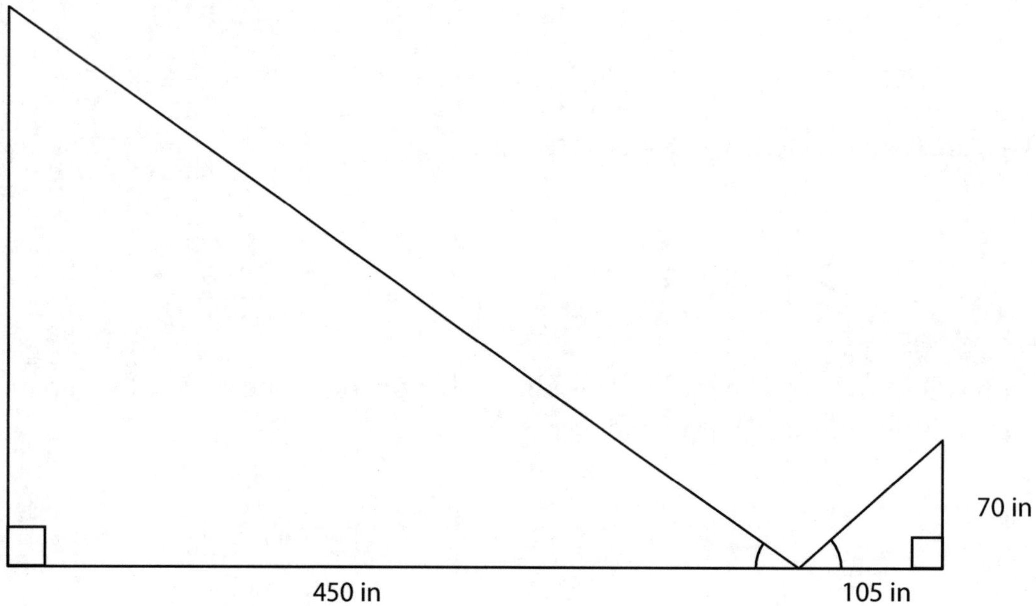

b. Are the triangles in the diagram similar? Explain.

The triangles in the diagram are similar because of the Angle-Angle (AA) Similarity Statement. This statement asserts that if two angles of one triangle are congruent to two angles of another triangle, then the triangles are similar.

c. What are the characteristics of similar triangles?

The angles of similar triangles are congruent and the sides are proportional.

d. Is it possible to determine the height of the house from the foundation to the gutters using the given information? Explain your answer.

Yes, it is possible to determine the height of the house from the foundation to the gutters using the given information. You can do so by creating a proportion using the two similar triangles.

e. What is the height of the house from the foundation to the gutters?

Create a proportion to find the height of the house from the foundation to the gutters.

$$\frac{\text{Blair's height}}{\text{height from foundation to gutters}} = \frac{\text{distance from mirror to Blair}}{\text{distance from mirror to foundation}}$$

$\frac{70}{x} = \frac{105}{450}$ Substitute.

$(70)(450) = (105)(x)$ Find the cross products.

$31{,}500 = 105x$ Simplify.

$x = 300$ Solve for x.

The height of the house from the foundation to the gutters is 300 inches.

f. How many 5-foot sections of downspout are needed for Blair's house?

Determine the length of the needed downspout in feet.

Divide the number of inches by 12.

length of needed downspout in feet = $300 \div 12 = 25$

The length of the needed downspout in feet is 25 feet.

Each section of downspout is 5 feet.

Divide the number of feet needed by 5 to determine the number of sections needed.

number of sections needed = $25 \div 5 = 5$

Blair needs 5 sections of downspout.

Common Core State Standard

G–SRT.4

Prove theorems about triangles. *Theorems include: a line parallel to one side of a triangle divides the other two sides proportionally, and conversely; the Pythagorean Theorem proved using triangle similarity.*

Problem-Based Task: Suddenly Sinking

In August 2012, a sinkhole in Louisiana swallowed part of the Earth's surface. This sudden erosion of land forced many people to evacuate as engineers worked to determine the effects of the sinkhole. Officials began their assessment of the area by first determining the size of the sinkhole. Surveyors worked to locate points near the sinkhole to calculate its diameter. The initial measurements are shown below. As days passed, the diameter of the sinkhole increased by 15%. What was the diameter of the sinkhole after the increase?

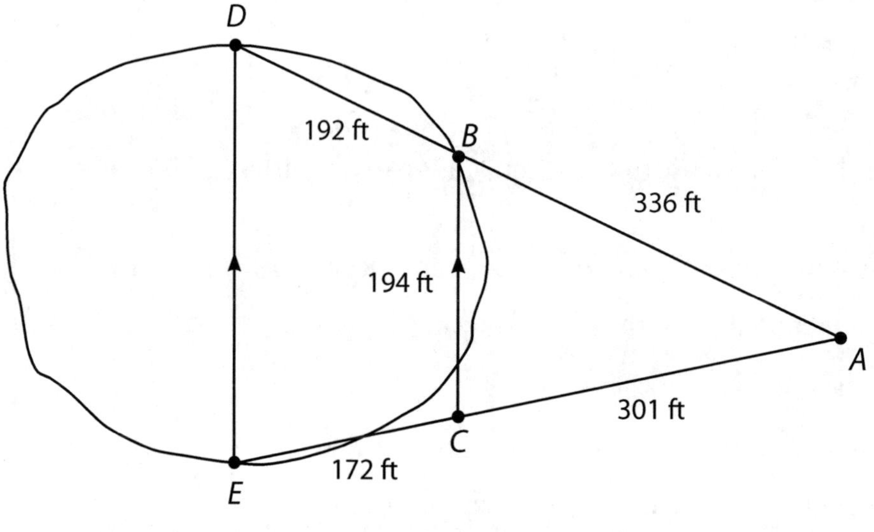

Problem-Based Task: Suddenly Sinking

Coaching

a. Are the triangles in the diagram similar? Explain.

b. What are the characteristics of similar triangles?

c. Is it possible to determine the diameter of the initial sinkhole using the given information? Explain your answer.

d. Determine the length of $\overline{DE}$.

e. What was the diameter of the initial sinkhole?

f. How are percent increases calculated?

g. What is 15% of the initial diameter?

h. What was the diameter after the sinkhole increased?

Problem-Based Task: Suddenly Sinking

Coaching Sample Responses

a. Are the triangles in the diagram similar? Explain.

The triangles in the diagram are similar. $\overline{BC}$ is parallel to $\overline{DE}$ of $\triangle ADE$.

According to the Triangle Proportionality Theorem, if a line parallel to one side of a triangle intersects the other two sides of the triangle, then the parallel line divides these two sides proportionally.

b. What are the characteristics of similar triangles?

The angles of similar triangles are congruent and the sides are proportional.

c. Is it possible to determine the diameter of the initial sinkhole using the given information? Explain your answer.

Yes, it is possible to determine the diameter of the initial sinkhole using the given information. You can do so by creating a proportion using the smaller similar triangle, $\triangle ABC$.

d. Determine the length of $\overline{DE}$.

Create a proportion to find the length of $\overline{DE}$.

$\frac{AD}{DE}=\frac{AB}{BC}$	Create a proportion.
$\frac{336+192}{x}=\frac{336}{194}$	Substitute known values.
$\frac{528}{x}=\frac{336}{194}$	Simplify.
$(528)(194)=(336)(x)$	Find the cross products.
$102{,}432=336x$	Simplify.
$x\approx 305$ feet	

The length of $\overline{DE}$ is approximately 305 feet.

e. What was the diameter of the initial sinkhole?

The diameter of the initial sinkhole was approximately 305 feet.

f. How are percent increases calculated?

One way to calculate a percent increase is to add the initial value to the product of the initial value and the decimal equivalent of the percent.

$$\text{initial value} + \frac{x}{100}(\text{initial value})$$

g. What is 15% of the initial diameter?

15% of 305 can be written numerically as 0.15(305).

The result is 45.75.

h. What was the diameter after the sinkhole increased?

Add 15% of the initial diameter to the initial diameter.

$$45.75 + 305 = 350.75$$

The diameter of the sinkhole after the increase was 350.75 feet.

G–SRT.4 • Geometry

Proving the Pythagorean Theorem Using Similarity

Common Core State Standard

G–SRT.4

Prove theorems about triangles. *Theorems include: a line parallel to one side of a triangle divides the other two sides proportionally, and conversely; the Pythagorean Theorem proved using triangle similarity.*

Problem-Based Task: Towering Heights

Cell towers are tall structures that provide service to customers with cell phones, smartphones, tablets, and other wireless communication devices. As the use of wireless devices increases, the need for cell towers goes up. According to the Federal Aviation Administration (FAA) and the Federal Communications Commission (FCC), any structure taller than 61 meters must be lit up so aircraft pilots can see it.

A cell phone company is interested in placing a 5-meter-tall antenna on an existing tower in order to boost their cell signal without having to build a new tower. A surveyor standing 7.5 meters from the base of the tower calculates the height of the existing tower. The line of sight from the lens on the surveyor's tripod to the base of the tower is 1 meter above the ground. If the 5-meter-tall antenna is added to the top of the tower, will the company be required to add lighting to the existing structure? Explain your reasoning.

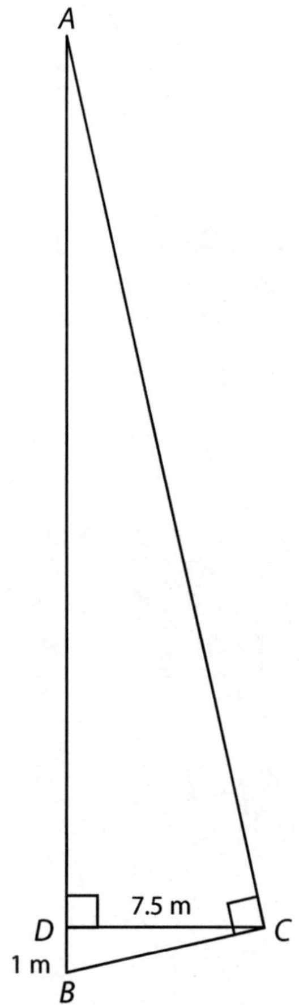

Problem-Based Task: Towering Heights

Coaching

a. Are the triangles in the diagram similar? Explain.

b. Write a statement indicating similar triangles.

c. What are the characteristics of similar triangles?

d. Is it possible to determine the height of the tower using the given information? Explain your answer.

e. Determine the length of $\overline{AD}$.

f. What is the height of the tower?

g. How tall will the tower be once the antenna is added?

h. Will the height of the tower with the antenna exceed the maximum height stated by the FAA and FCC?

i. Will the company be required to add lighting to the tower once the antenna is added?

Problem-Based Task: Towering Heights

Coaching Sample Responses

a. Are the triangles in the diagram similar? Explain.

Yes, the triangles are similar. The triangle created from the base of the tower to the height of the tower to the surveyor is a right triangle. The segment from the 90° angle to the tower is perpendicular to the tower and is the altitude of the triangle. By definition, the altitude of the right triangle creates two smaller, similar right triangles.

b. Write a statement indicating similar triangles.

$\triangle ABC \sim \triangle CDB \sim \triangle ADC$

c. What are the characteristics of similar triangles?

The angles of similar triangles are congruent and the sides are proportional.

d. Is it possible to determine the height of the tower using the given information? Explain your answer.

Yes, it is possible to determine the height of the tower using the given information by creating a proportion using the two smaller similar triangles, $\triangle CDB$ and $\triangle ADC$.

e. Determine the length of $\overline{AD}$.

Create a proportion to find the length of $\overline{AD}$.

$$\frac{CD}{AD} = \frac{BD}{CD}$$

$$\frac{7.5}{x} = \frac{1}{7.5}$$ Substitute known values.

$(7.5)(7.5) = (1)(x)$ Find the cross products.

$56.25 = x$ Simplify.

The length of $\overline{AD}$ is 56.25 meters.

f. What is the height of the tower?

To find the height of the tower, add the length of $\overline{AD}$ to the length of $\overline{BD}$.

$AD + BD = 56.25 + 1 = 57.25$

The height of the tower is 57.25 meters.

g. How tall will the tower be once the antenna is added?

Add the height of the antenna to the height of the tower.

$5 + 57.25 = 62.25$

The tower will be 62.25 meters tall after the antenna is added.

h. Will the height of the tower with the antenna exceed the maximum height stated by the FAA and FCC?

Yes; 62.25 is greater than 61 meters, the maximum height stated by the FAA and FCC.

i. Will the company be required to add lighting to the tower once the antenna is added?

Yes; the tower with the antenna will exceed 61 meters, so lighting will have to be added.

Common Core State Standard

G–SRT.5

Use congruence and similarity criteria for triangles to solve problems and to prove relationships in geometric figures.

Problem-Based Task: Too Tall?

Parks directors routinely assess the health of the trees in recreation areas. If trees are found to be diseased, they are often treated. If trees become too weak, they are removed before they become a danger to people and structures. Gorge Park is a rectangular park measuring 400 feet by 200 feet and is enclosed by a fence. A diseased tree needing removal stands in the center of the park. Tree removers must avoid having the tree fall on the fence. If necessary, the tree can be trimmed prior to being cut down. The 6-foot-tall parks director measured the length of the shadow cast by the tree to be 147 feet and the length of his own shadow to be 9 feet. Does the tree's trunk need to be trimmed prior to cutting it down to avoid damaging the fence?

Problem-Based Task: Too Tall?

Coaching

a. Draw triangles to represent the height of the parks director, the diseased tree, and their shadows.

b. Are the triangles in your diagram similar? Explain.

c. What are the characteristics of similar triangles?

d. Is it possible to determine the height of the diseased tree using the given information? Explain your answer.

e. What is the height of the diseased tree?

f. Draw and label a diagram of Gorge Park and mark the location of the diseased tree.

g. Compare the height of the tree to the distance between the tree and each fence.

h. Does the tree need to be trimmed prior to cutting it down to avoid damaging the fence?

Problem-Based Task: Too Tall?

Coaching Sample Responses

a. Draw triangles to represent the height of the parks director, the diseased tree, and their shadows.

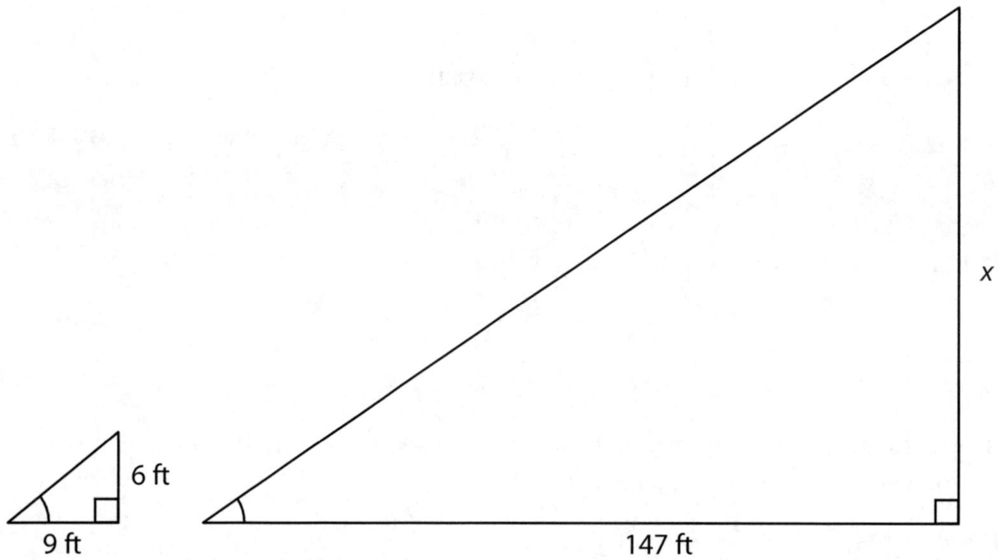

b. Are the triangles in your diagram similar? Explain.

The triangles in the diagram are similar because of the Angle-Angle (AA) Similarity Statement, which states that if two angles of one triangle are congruent to two angles of another triangle, then the triangles are similar.

c. What are the characteristics of similar triangles?

The angles of similar triangles are congruent and the sides are proportional.

d. Is it possible to determine the height of the diseased tree using the given information? Explain your answer.

Yes, it is possible to determine the height of the diseased tree using the given information by creating a proportion using the two similar triangles.

e. What is the height of the diseased tree?

Create a proportion to find the height of the diseased tree.

$$\frac{\text{height of the director}}{\text{height of the tree}} = \frac{\text{length of the director's shadow}}{\text{length of the tree's shadow}}$$

$\frac{6}{x} = \frac{9}{147}$	Substitute known values.
$(6)(147) = (9)(x)$	Find the cross products.
$882 = 9x$	Simplify.
$x = 98$	Solve for x.

The height of the diseased tree is 98 feet.

f. Draw and label a diagram of Gorge Park and mark the location of the diseased tree.

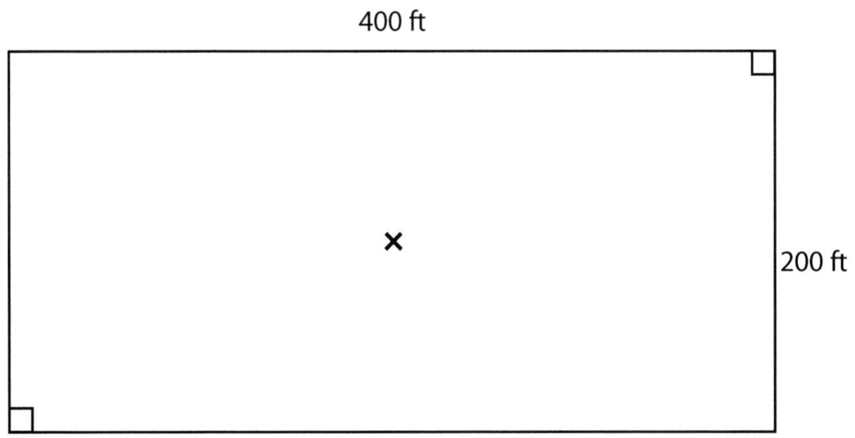

g. Compare the height of the tree to the distance between the tree and each fence.

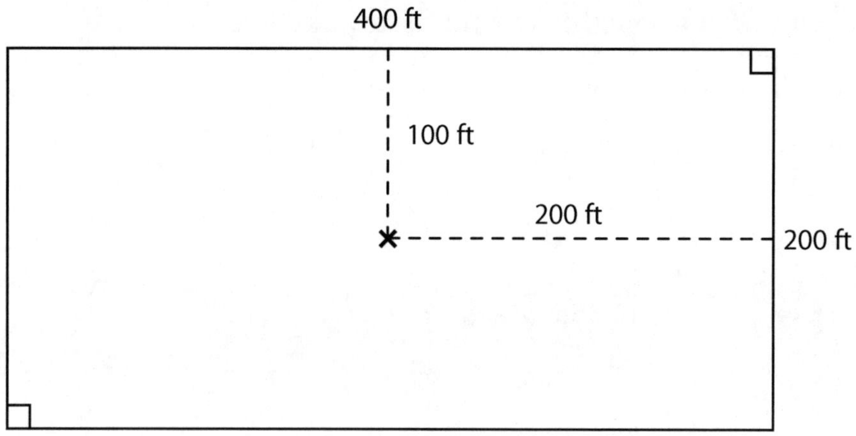

The diseased tree is located in the center of the park. According to the diagram, there are 200 feet of park to the east and west of the tree and 100 feet of park north and south of the tree. The height of the tree is less than both of these distances to the fence.

h. Does the tree need to be trimmed prior to cutting it down to avoid damaging the fence?

The tree does not need to be trimmed. If cut properly, the tree will not fall on the fence.

NAME:

G–SRT.6 • Geometry

Defining Trigonometric Ratios

Common Core State Standard

G–SRT.6

Understand that by similarity, side ratios in right triangles are properties of the angles in the triangle, leading to definitions of trigonometric ratios for acute angles.

Problem-Based Task: Building a Porch Ramp

Bill's father is building a ramp to reach the front porch of their house. The ramp will rise from the flat ground to the top of the porch. The porch is 2 feet tall. The ramp will rise at a 10° angle from the ground. How long does the ramp need to be? What is the sine of 10°?

Problem-Based Task: Building a Porch Ramp

Coaching

a. Create a drawing of the problem.

b. Is there a triangle in this problem?

c. Do you have enough information to draw a similar triangle to the drawing?

d. Can you make a precise drawing to scale that shows the known length and angles?

e. Is your small triangle similar to the large scale of the ramp?

f. What is the scale factor between the height of the porch in your drawing and the porch's actual height?

g. Which length in your drawing corresponds to the ramp length?

h. How do you use the length in the drawing and the scale factor to find the length of the ramp?

i. What is the ratio for the sine of 10°?

Problem-Based Task: Building a Porch Ramp

Coaching Sample Responses

a. Create a drawing of the problem.

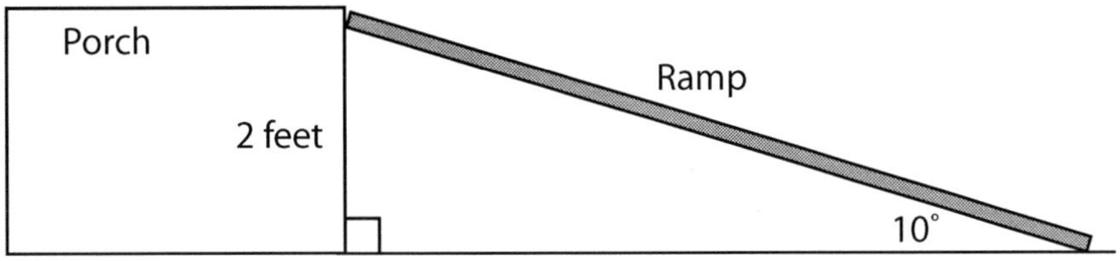

b. Is there a triangle in this problem?

Yes; the triangle is formed by the ramp (the hypotenuse), the height of the porch, and the flat ground.

c. Do you have enough information to draw a similar triangle?

Yes. There is information for two angles: an angle of 10° is given, and the angle the ground makes with the porch is 90°. This means you can draw a similar triangle to the triangle created by the ramp. The third angle is 90° – 10° = 80°.

d. Can you make a precise drawing to scale that shows the known length and angle?

Yes. You can use a protractor to create a 10° angle and a 90° angle, and a ruler to draw a side length that corresponds in units to the porch height (suggested scale: 1/2 inch = 1 foot). Scale factors of 1 inch : 1 foot will require paper that is at least 12 inches long.

e. Is your small triangle similar to the large scale of the ramp?

Yes. This is a scaled drawing that includes the 10° angle, the 90° angle, and the height of the porch.

f. What is the scale factor between the height of the porch in your drawing and the porch's actual height?

The scale factor will vary depending on how large the triangle is in the drawing. If students use the ratio 1/2 : 1, the scale factor is 1 : 24, making $k = 24$.

g. Which length in your drawing corresponds to the ramp length?

The length is the hypotenuse on the drawing.

h. How do you use the length in the drawing and the scale factor to find the length of the ramp?

Multiply the length of the hypotenuse by the scale factor.

First, measure the hypotenuse. If the scale on the drawing is 1/2 : 1, the length of the ramp in the drawing would equal approximately 5.75 inches.

Multiply the length of the drawn ramp by the scale factor to determine the length of the actual ramp.

$5.75 \bullet 24 = 138$ inches = 11.5 feet

Compare students' answers and see the variation in measurement data. Almost all the answers should range from 10.75 feet to 12.25 feet.

i. What is the sine of 10°?

$$\sin 10^\circ = \frac{2}{11.5} \approx 0.174$$

G–SRT.7 • Geometry

Exploring Sine and Cosine As Complements

Common Core State Standard

G–SRT.7

Explain and use the relationship between the sine and cosine of complementary angles.

Problem-Based Task: School Vegetable Garden

A high school biology class is planning a vegetable garden. The garden will be on the west side of the school building. The students are concerned that the garden will not get enough direct sunlight in the morning if the building's shadow shades the garden too much. Starting May 1, the garden needs direct sunlight by 10 A.M. at the latest. The school building is 8 meters high. The garden is 6 meters from the building. The students find on the Internet that the angle of the sun's rays at 10 A.M. on May 1 will be 60° above the horizon. How long will the building's shadow be? Is the chosen location a good spot for the garden? What are the sine of 30° and the cosine of 60°?

Problem-Based Task: School Vegetable Garden

Coaching

a. Draw a picture of the problem.

b. What are the measures of the angles in the drawing's triangle?

c. What is the ratio between the shortest side length and the hypotenuse in this triangle?

d. What is the length of the hypotenuse if the shortest side length of this triangle is 1 inch?

e. Find the length of the longer side by using the Pythagorean Theorem.

f. What are the basic side lengths of such a triangle if the shortest side length is 1 unit?

g. What is the scale factor of the triangle in the drawing?

h. How can you use the scale factor to find the length of the building's shadow?

i. Is the chosen location a good spot for the garden? Explain.

j. What are the sine of 30° and the cosine of 60°?

Problem-Based Task: School Vegetable Garden

Coaching Sample Responses

a. Draw a picture of the problem.

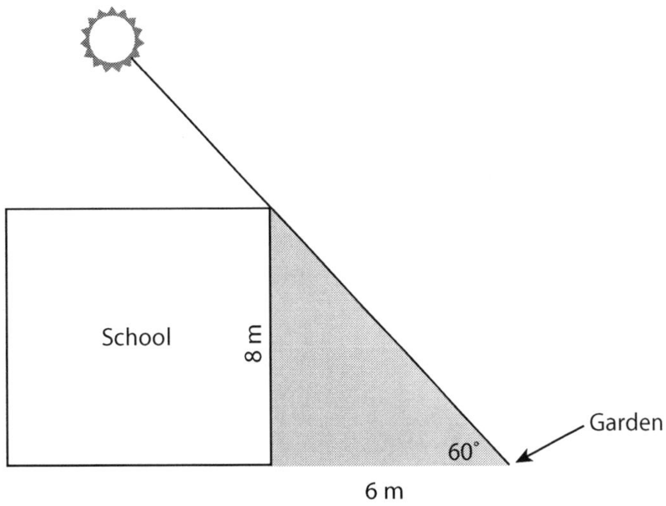

b. What are the measures of the angles in the drawing's triangle?

The angles measure 30°, 60°, and 90°.

c. What is the ratio between the shortest side length and the hypotenuse in this triangle?

Draw a triangle that has angles of 30°, 60°, and 90°.

Measure the shortest side length, *a*.

Measure the hypotenuse, *c*.

Divide the hypotenuse length by the shortest side length, *a*.

The ratio is 2 : 1 (or 1 : 2). The hypotenuse is twice as long as the shortest leg.

d. What is the length of the hypotenuse if the shortest side length of this triangle is 1 inch?

Draw a 30°–60°–90° triangle so that the shortest side length = 1 inch.

The length of the hypotenuse = 2 inches.

e. Find the length of the longer side by using the Pythagorean Theorem.

Substitute values for *a* and *c*.

$1^2 + b^2 = 2^2$

$b^2 = 3$

$b = \sqrt{3}$ inches

f. What are the basic side lengths of a 30°–60°–90° triangle if the shortest side length is 1 unit?

The basic proportional lengths opposite the angles in a 30°–60°–90° triangle are:

opposite side length of 30° angle (shortest side length): 1

opposite side length of 60° angle (longest side length): $\sqrt{3}$

opposite side length of 90° angle (hypotenuse): 2

g. What is the scale factor, *k*, of the 30°–60°–90° triangle in the drawing?

The only known side length (the longer side opposite the 60° angle) is 8 meters.

The basic length of the longer side in a 30°–60°–90° triangle is $\sqrt{3}$.

$$\sqrt{3} \bullet k = 8 \text{ meters}$$
$$k = \frac{8}{\sqrt{3}}$$
$$k \approx 4.619 \text{ meters}$$

h. How can you use the scale factor to find the length of the building's shadow?

The shadow is opposite the 30° angle in the drawing. This makes it the shortest side length.

The basic length of the shortest side in a 30°–60°–90° triangle is 1.

Multiply the scale factor by the basic side length of the triangle that corresponds with the building's shadow. This is the short side of the triangle, and the basic length is 1. Therefore, the length of the building's shadow is approximately 4.619 meters.

i. Is the chosen location a good spot for the garden? Explain.

Yes, because the garden will not have shade at 10 A.M. on May 1. Since the garden will be placed 6 meters from the building, and the building's shadow will be about 4.619 meters long, the garden will be outside of the shade.

j. What are the sine of 30° and the cosine of 60°?

$\sin 30° = 0.5$

$\cos 60° = 0.5$

G–SRT.8 • Geometry

Calculating Sine, Cosine, and Tangent

Common Core State Standard

G–SRT.8

Use trigonometric ratios and the Pythagorean Theorem to solve right triangles in applied problems.★

Problem-Based Task: Replanting Project

The environmental club at your school has taken on the task of planting a garden to provide fresh fruit, herbs, and vegetables to the school cafeteria. The approved area has several trees that need to be moved and replanted so that you can use the garden space more efficiently. A building supply company donates several pieces of wire, each 13 feet in length, to help brace the replanted trees.

As the president of the environmental club, you must provide the specifications to the club members and volunteers before they can begin the replanting process. At most, the ladders you have can reach 11 feet 3 inches up the trunk of a tree. Given this information, what angle should the wires make with the ground, and at what horizontal distance from the tree trunk should the wires be staked to the ground?

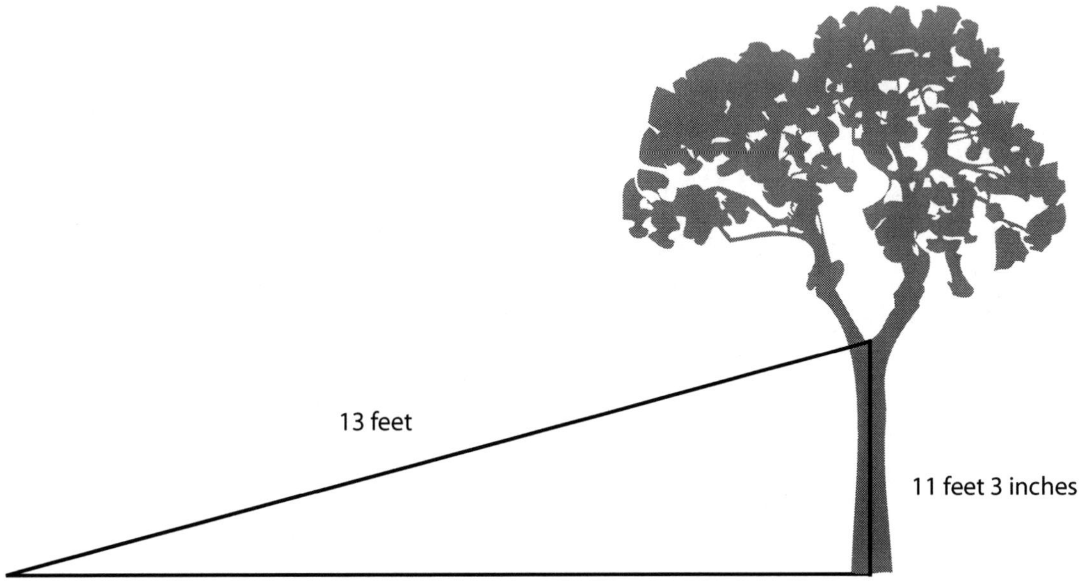

Problem-Based Task: Replanting Project

Coaching

a. What is the maximum height, in inches, each wire can be secured to a tree?

b. What is the length of each wire in inches?

c. Based on the given information, which trigonometric function is required to find the angle at which each wire should be staked to the ground? What is the angle measure?

d. What right triangle property should you use to most accurately find the distance from the trunk of the tree to the spot where the wire will be staked to the ground?

e. What is the distance, in inches, from the trunk of the tree to the spot where the wire will be staked to the ground?

Problem-Based Task: Replanting Project

Coaching Sample Responses

a. What is the maximum height, in inches, each wire can be secured to a tree?

The wires can be secured as high up as the ladders will reach. Since this is known to be 11 feet 3 inches, start by converting 11 feet to inches.

$$(11\text{ ft})\left(\frac{12\text{ in}}{1\text{ ft}}\right)=132\text{ in}$$

Add 3 inches to determine the total height in inches.

$$132 + 3 = 135$$

The maximum height each wire can be secured is 135 inches.

b. What is the length of each wire in inches?

Convert the known length of the wires (13 feet) to inches.

$$(13\text{ ft})\left(\frac{12\text{ in}}{1\text{ ft}}\right)=156\text{ in}$$

c. Based on the given information, which trigonometric function is required to find the angle at which each wire should be staked to the ground? What is the angle measure?

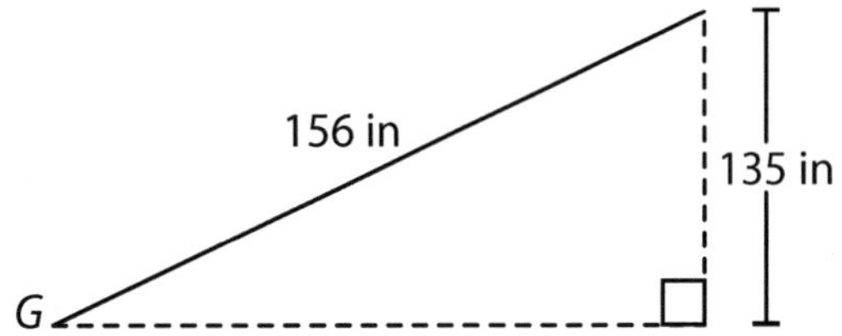

The opposite side of the angle and the hypotenuse are known, so sine should be used.

$$\sin G = \frac{135}{156}$$

$$G = \sin^{-1}\left(\frac{135}{156}\right)$$

$G = 59.926°$. The angle of elevation is approximately 60°.

d. What right triangle property should you use to most accurately find the distance from the trunk of the tree to the spot where the wire will be staked to the ground?

Since two side lengths were given, you could use either trigonometry or the Pythagorean Theorem. However, in an effort to be as precise as possible, the best choice is the Pythagorean Theorem. With trigonometry, you would have to use the acute angle *G* that we just approximated in part c.

e. What is the distance, in inches, from the trunk of the tree to the spot where the wire will be staked to the ground?

From part d we determined that the Pythagorean Theorem is the most accurate method for determining the distance.

d = unknown

leg = 135 in

hypotenuse = 156 in

$d^2 + 135^2 = 156^2$

$d^2 + 18{,}225 = 24{,}336$

$d^2 = 6111$

$d = \pm\sqrt{6111}$

$d = 78.173$

The distance from the tree trunk to the staked wire is about 78 inches.

G–SRT.8 • Geometry

Calculating Cosecant, Secant, and Cotangent

Common Core State Standard

G–SRT.8

Use trigonometric ratios and the Pythagorean Theorem to solve right triangles in applied problems.★

Problem-Based Task: A Baseball Triangle

Three baseball players are in the outfield at locations *A*, *B*, and *C*, as shown in the diagram below. The distance between *A* and *B* is the same as the distance between *A* and *C*. Their positions form an isosceles triangle. How could you use the properties of isosceles triangles to find the cosecant, secant, and tangent of $\angle C$? Calculate the cosecant, secant, and tangent of $\angle C$, and then find the angle measures.

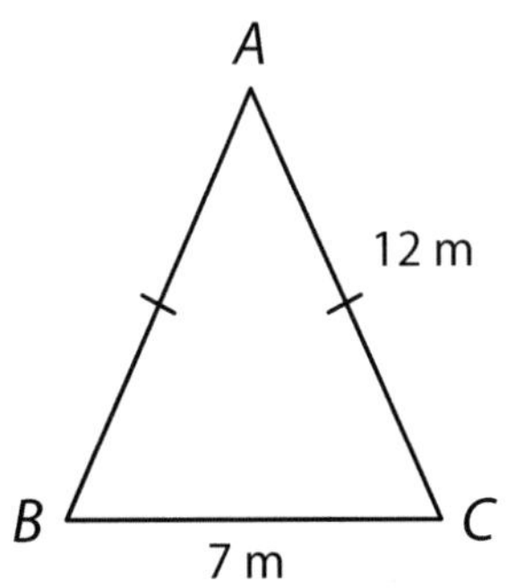

Problem-Based Task: A Baseball Triangle

Coaching

a. What type of triangle is needed in order to calculate with trigonometry?

b. How could you create a right triangle using the given isosceles triangle?

c. What does the altitude, or height, of an isosceles triangle create with the base of the triangle? What does that mean about the base of the isosceles triangle?

d. What information do you need to calculate the cosecant, secant, and tangent of $\angle C$?

e. Calculate the cosecant, secant, and tangent of $\angle C$.

f. How did you calculate the cosecant, secant, and tangent of $\angle C$?

g. What buttons do you need to use on your calculator to find the angle measures given a trigonometric ratio?

h. What is the relationship between the buttons you have to use and the trigonometric ratios you calculated?

i. What are the measures of $\angle A$, $\angle B$, and $\angle C$?

Problem-Based Task: A Baseball Triangle

Coaching Sample Responses

a. What type of triangle is needed in order to calculate with trigonometry?

A right triangle is needed in order to use any trigonometric functions.

b. How could you create a right triangle using the given isosceles triangle?

Draw the height, or the altitude, of the triangle.

c. What does the altitude of an isosceles triangle create with the base of the triangle? What does that mean about the base of the isosceles triangle?

The altitude of an isosceles triangle is the perpendicular bisector of the base. That means the right triangles created will each have a short leg of 3.5 meters since the base of the isosceles triangle is 7 meters.

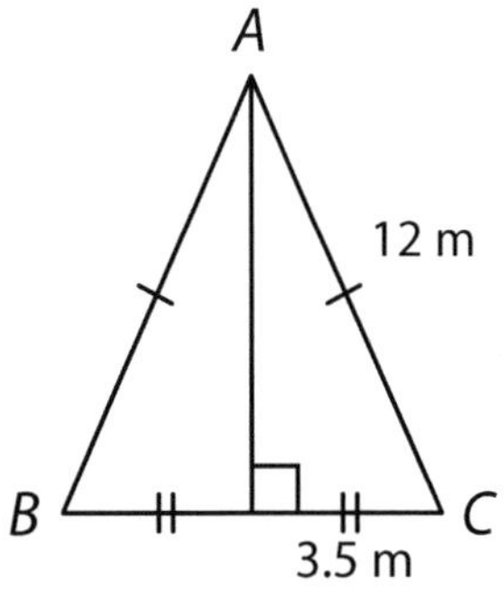

d. What information do you need to calculate the cosecant, secant, and tangent of $\angle C$?

Recall that:

$$\csc\theta = \frac{\text{hypotenuse}}{\text{opposite}} \qquad \sec\theta = \frac{\text{hypotenuse}}{\text{adjacent}} \qquad \cot\theta = \frac{\text{adjacent}}{\text{opposite}}$$

Since the side opposite from $\angle C$ is not known and is needed in order to calculate both cosecant and cotangent, use the Pythagorean Theorem to determine the opposite side. Let *h*, height, stand for the opposite side.

$3.5^2 + h^2 = 12^2$

$12.25 + h^2 = 144$

$h^2 = 131.75$

$h = \sqrt{131.75}$

$h = 11.478$

The opposite side, or the height, *h*, is 11.478 meters.

e. Calculate the cosecant, secant, and tangent of $\angle C$.

$$\csc C = \frac{\text{hypotenuse}}{\text{opposite}} = \frac{12}{11.478} \approx 1.045$$

$$\sec C = \frac{\text{hypotenuse}}{\text{adjacent}} = \frac{12}{3.5} \approx 3.429$$

$$\cot C = \frac{\text{adjacent}}{\text{opposite}} = \frac{3.5}{11.478} \approx 0.305$$

f. How did you calculate the cosecant, secant, and tangent of $\angle C$?

After creating a right triangle using the altitude of the isosceles triangle, use the Pythagorean Theorem to find the measure of the side opposite from $\angle C$. Given that side length, set up the trigonometric ratio for each reciprocal function and divide their respective ratios to get the decimal approximation.

g. What buttons do you need to use on your calculator to find the angle measures given a trigonometric ratio?

You must use the sine (SIN), cosine (COS), and tangent (TAN) buttons because there are no buttons for cosecant (CSC), secant (SEC), and cotangent (COT).

h. What is the relationship between the functions of the buttons you have to use and the trigonometric ratios you calculated?

The cosecant is the reciprocal of sine, the secant is the reciprocal of cosine, and the cotangent is the reciprocal of tangent.

i. What are the measures of $\angle A$, $\angle B$, and $\angle C$?

Use the following relationships and find the reciprocal of each trigonometric ratio.

$\csc C = \frac{\text{hypotenuse}}{\text{opposite}} = \frac{12}{11.478} \approx 1.045$ becomes $\sin C = \frac{\text{opposite}}{\text{hypotenuse}} = \frac{11.478}{12} \approx 0.957$.

$\sec C = \frac{\text{hypotenuse}}{\text{adjacent}} = \frac{12}{3.5} \approx 3.429$ becomes $\cos C = \frac{\text{adjacent}}{\text{hypotenuse}} = \frac{3.5}{12} \approx 0.292$.

$\cot C = \frac{\text{adjacent}}{\text{opposite}} = \frac{3.5}{11.478} \approx 0.305$ becomes $\tan C = \frac{\text{opposite}}{\text{adjacent}} = \frac{11.478}{3.5} \approx 3.279$.

Now, take the inverse of at least one of the trigonometric functions to find the measure of $\angle C$. Verify with the other two trigonometric ratios that the angle measure is the same for any trigonometric ratio used.

$$\sin C = \frac{\text{opposite}}{\text{hypotenuse}} = \frac{11.478}{12} \approx 0.957$$

$$\sin C \approx 0.957$$

$$C \approx \sin^{-1}(0.957) \approx 73°$$

$$\cos C = \frac{\text{adjacent}}{\text{hypotenuse}} = \frac{3.5}{12} \approx 0.292$$

$$\cos C \approx 0.292$$

$$C \approx \cos^{-1}(0.292) \approx 73°$$

$$\tan C = \frac{\text{opposite}}{\text{adjacent}} = \frac{11.478}{3.5} \approx 3.279$$

$$\tan C \approx 3.279$$

$$C \approx \tan^{-1}(3.279) \approx 73°$$

We know that $\triangle ABC$ is an isosceles triangle and the angles opposite the congruent sides are congruent. Therefore, $\angle C \cong \angle B$, which means that the measure of $\angle B$ is also 73°. Use the Triangle Sum Theorem to find the measure of the third angle: 180 – (73 + 73) = 34. The measure of $\angle A$ is 34°.

Common Core State Standard

G–SRT.8

Use trigonometric ratios and the Pythagorean Theorem to solve right triangles in applied problems.★

Problem-Based Task: Fighting Flames from a Distance

You are a firefighter on call at a burning building. Your colleagues are on the roof preparing to help put out the blaze by entering the building through the third-floor window. They need you to find the distance from the roof to the windowsill and then determine if the firefighters on the ground are close enough to the building for the water to reach the flames through the window on the third floor.

You observe from below. The angle of elevation to the windowsill is 18° and the angle of elevation to the top of the building is 31°. You are standing 65 feet away from the building and your eyes are 5 feet above the ground, as shown in the diagram. You hold the hose at eye level in order to take aim at the third-floor window.

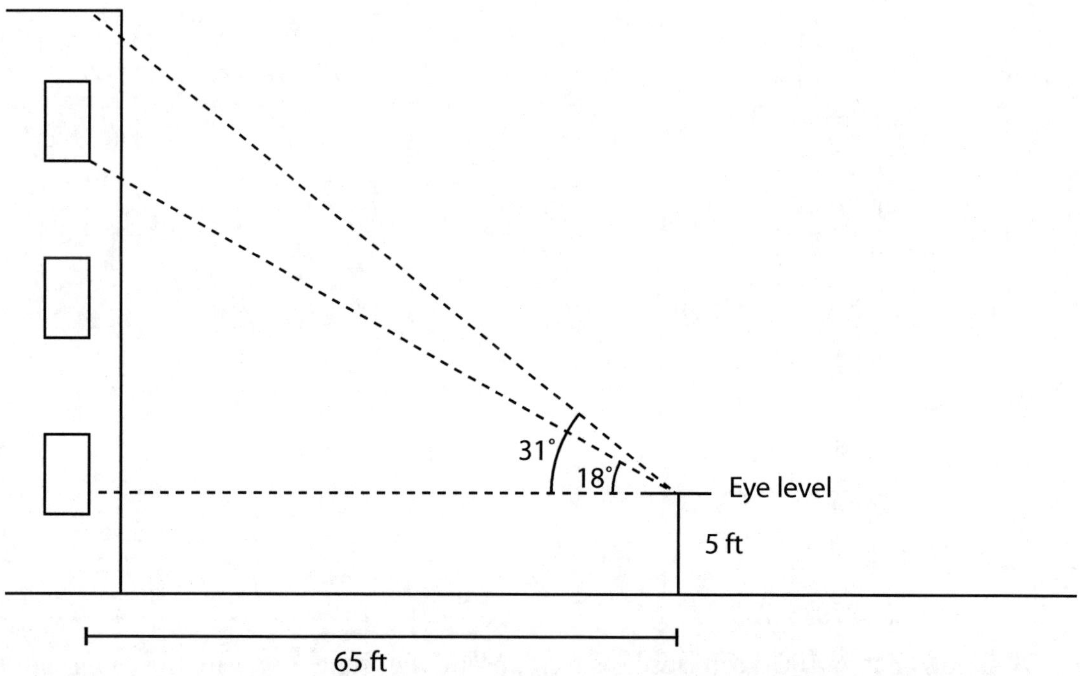

To the nearest foot, what is the distance from the roof to the windowsill that the firefighters will need to descend by rope to enter the building? If the hoses can spray water at a distance of 100 feet, are the firefighters standing close enough to the building to put the flames out at the third-floor window?

Problem-Based Task: Fighting Flames from a Distance

Coaching

a. In order to determine the distance from the windowsill to the roof, what other information do you need to find first?

b. What is the vertical height from the roof to your eye level? Sketch the triangle that provides the information necessary to answer the question.

c. What is the vertical height from the windowsill to your eye level?

d. What is the distance from the roof to the third-floor windowsill?

e. How can you use the information you've already found to determine the distance the water has to travel?

f. What distance does the water have to travel to reach the window?

g. Are the firefighters on the ground close enough to the building?

Problem-Based Task: Fighting Flames from a Distance

Coaching Sample Responses

a. In order to determine the distance from the windowsill to the roof, what other information do you need to find first?

Since the vertical height from the roof to the windowsill is not a side of a right triangle, you have to use the two right triangles that are given to find the distance from your eye level to the windowsill and the distance from your eye level to the roof. Subtracting the two values will yield the distance from the roof to the windowsill.

b. What is the vertical height from the roof to your eye level? Sketch the triangle that provides the information necessary to answer the question.

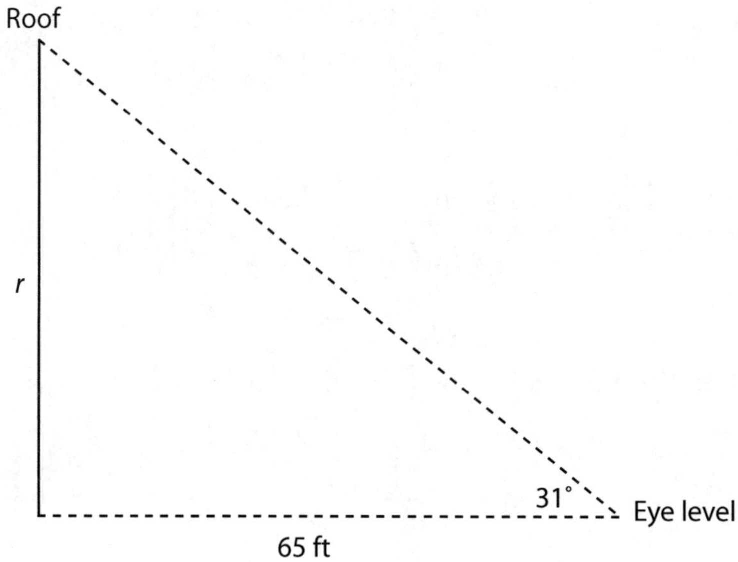

Let *r* represent the vertical height from the roof to your eye level. Given an angle of elevation of 31°, the leg adjacent to the angle of elevation is 65 feet. To calculate the leg opposite the angle of elevation, the tangent function is required.

$$\tan 31° = \frac{r}{65}$$

$$65 \bullet \tan 31° = r$$

On a TI-83/84:

Step 1: Press [65][TAN][31][)].

Step 2: Press [ENTER].

$r \approx 39.056$

On a TI-Nspire:

Step 1: In the calculate window from the home screen, enter 65, then press [trig]. Use the keypad to select "tan," then enter 31.

Step 2: Press [enter].

$r \approx 39.056$

The vertical height from the roof to your eye level is about 39 meters.

c. What is the vertical height from the windowsill to your eye level?

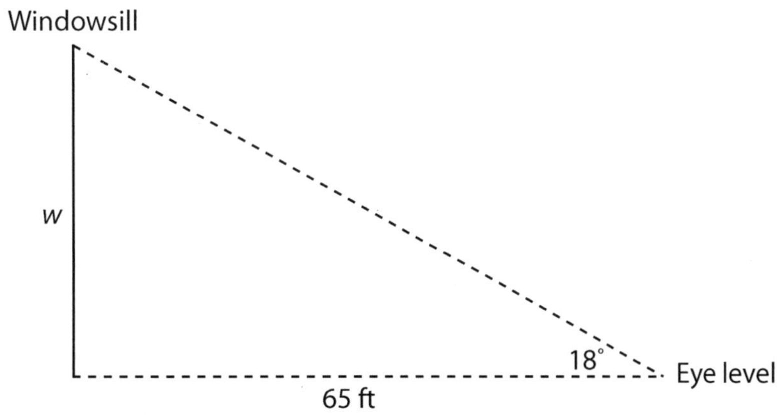

Let w represent the vertical height from the windowsill to your eye level. Given an angle of elevation of 18°, the leg adjacent to the angle of elevation is 65 feet long. To calculate the leg opposite the angle of elevation, the tangent function is required.

$$\tan 18° = \frac{w}{65}$$

$$65 \bullet \tan 18° = w$$

On a TI-83/84:

Step 1: Press [65][TAN][18][)].

Step 2: Press [ENTER].

$w \approx 21.120$

On a TI-Nspire:

Step 1: In the calculate window from the home screen, enter 65, then press [trig]. Use the keypad to select "tan," then enter 18.

Step 2: Press [enter].

$w \approx 21.120$

The vertical height from the windowsill to your eye level is about 21 feet.

d. What is the distance from the roof to the third-floor windowsill?

To find the distance from the roof to the windowsill, subtract 21.120 from 39.056.

$39.056 - 21.120 = 17.936$

The distance is approximately 18 feet.

e. How can you use the information you've already found to determine the distance the water has to travel?

Using the diagram in part c, use the Pythagorean Theorem to find the hypotenuse, or the distance the water will travel from the spout of the hose at eye level to the windowsill.

f. What distance does the water have to travel to reach the window?

$21.120^2 + 65^2 = d^2$

$446.054 + 4225 = d^2$

$4671.054 = d^2$

$\pm\sqrt{4671.054} = d$

$d = 68.345$

The distance the water has to travel to reach the window is about 68 feet.

g. Are the firefighters on the ground close enough to the building?

Yes, the firefighters on the ground are close enough to the building because the water sprays 100 feet and the distance the water needs to travel is about 68 feet.

G–C.1; G–C.2 • Geometry

Similar Circles and Central and Inscribed Angles

Common Core State Standards

G–C.1

Prove that all circles are similar.

G–C.2

Identify and describe relationships among inscribed angles, radii, and chords. *Include the relationship between central, inscribed, and circumscribed angles; inscribed angles on a diameter are right angles; the radius of a circle is perpendicular to the tangent where the radius intersects the circle.*

Problem-Based Task: Following in Archimedes' Footsteps

The precise determination of the value of *pi* was of interest to the ancient Greeks for many years. Several approximations were calculated, but Archimedes was credited with determining a very precise approximation of *pi*. Identify several circular objects in your classroom and verify the approximation of a circle's circumference to its diameter.

Problem-Based Task: Following in Archimedes' Footsteps

Coaching

a. How could you use string and a ruler to measure the circumferences of circular figures?

b. What should you do if the circular object does not have a clearly defined center?

c. If you are approximating the center, what does that tell you about the final average you will find?

d. Locate several circular objects and measure the circumference and diameter of each. Calculate the ratio of the circle's circumference to its diameter. Use the table to record your information.

Circular object	Circumference (cm)	Diameter (cm)	$\frac{C}{d}$

e. Will the number of circular objects that you choose have an effect on your approximation?

f. What could your class do to arrive at a better approximation?

g. What will the average help you find?

h. How does what you have found relate to the idea of circle similarity?

Problem-Based Task: Following in Archimedes' Footsteps

Coaching Sample Responses

a. How could you use string and a ruler to measure the circumferences of circular figures?

Wrap the string around the figure and use your finger as a marker for the start and end of the circle after one revolution. Then use the ruler to measure the length of the string that mapped the circumference of the figure.

b. What should you do if the circular object does not have a clearly defined center?

Mark a center using an agreed-upon point.

c. If you are approximating the center, what does that tell you about the final average you will find?

If you approximate, your final answer will be less precise.

d. Locate several circular objects and measure the circumference and diameter of each. Calculate the ratio of the circle's circumference to its diameter. Use the table to record your information.

e. Will the number of circular objects that you choose have an effect on your approximation?

Yes; the more circular objects that are used, the more accurate the approximation.

f. What could your class do to arrive at a better approximation?

The class could, after collecting averages in small groups, find the average for the entire class.

g. What will the average help you find?

The average is a measure of the middle of a data set. In this case, it will give us one relationship of $\frac{C}{d}$ that is representative of the entire set.

h. How does what you have found relate to the idea of circle similarity?

All students should find a value close to 3.14, a constant for all circles, thus realizing that all circles are similar.

Common Core State Standard

G–C.2

Identify and describe relationships among inscribed angles, radii, and chords. *Include the relationship between central, inscribed, and circumscribed angles; inscribed angles on a diameter are right angles; the radius of a circle is perpendicular to the tangent where the radius intersects the circle.*

Problem-Based Task: Masking the Problem

In art class, you are creating a mask from a circular plate. The circumference of the plate is 30 centimeters and you have been instructed to cut out the central angle such that its intercepted major arc measures 200°. What is the length of the arc on the angle that you need to cut out?

Problem-Based Task: Masking the Problem

Coaching

a. What is the relationship between the central angle and its intercepted arc?

b. What is the measure of a circle?

c. What equation could you set up to find the length of the arc?

d. What is the length of the arc?

Problem-Based Task: Masking the Problem

Coaching Sample Responses

a. What is the relationship between the central angle and its intercepted arc?

A central angle is congruent to its intercepted arc.

b. What is the measure of a circle?

A circle has a measure of 360°.

c. What equation could you set up to find the length of the arc?

You could set up a proportion relating the degrees of a circle and the arc angle to the circumference of the circle and the arc length.

d. What is the length of the arc?

$$\frac{200}{360}=\frac{x}{30}$$

$$6000 = 360x$$

$$x=16\frac{2}{3}$$

The length of the arc should be about 16.7 centimeters.

G–C.2 • Geometry

Properties of Tangents of a Circle

Common Core State Standard

G–C.2

Identify and describe relationships among inscribed angles, radii, and chords. *Include the relationship between central, inscribed, and circumscribed angles; inscribed angles on a diameter are right angles; the radius of a circle is perpendicular to the tangent where the radius intersects the circle.*

Problem-Based Task: The Circus Is in Town! Is It Safe?

The circus has just arrived in town and is setting up. The fire inspector is on site to make sure that the circus meets local fire codes. The fire code requires that the circus have at least 175 feet of exterior walkways to pass inspection so that audience members are safe.

In the diagram below, all lines that appear to be tangent are tangent. The distance from the ticket booth, exit A, to the center of the concession stand is 25 feet. It is the same distance from the center of the concession stand to the petting zoo, at exit B. Use this information and the diagram to determine if the circus will pass inspection.

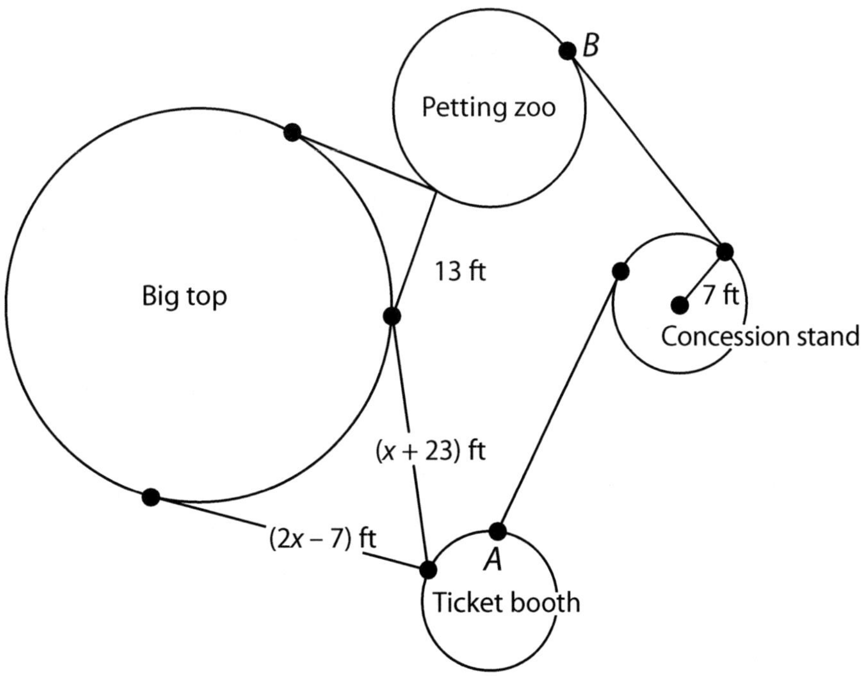

Problem-Based Task: The Circus Is in Town! Is It Safe?

Coaching

a. What is the length of the walkway tangent to the concession stand, which starts at the ticket booth?

b. What is the length of the walkway tangent to the concession stand, which ends at the petting zoo?

c. What is the length of the walkway tangent to the big top, which extends from the same point as the walkway that is 13 feet long?

d. What is the value for *x* for the two walkways tangent to the big top, extending from the same point at the ticket booth?

e. What is the length of the lines tangent to the big top, extending from the same point at the ticket booth?

f. What is the sum of the lengths of all of the exterior walkways?

g. Will the circus pass inspection?

Problem-Based Task: The Circus Is in Town! Is It Safe?

Coaching Sample Responses

a. What is the length of the walkway tangent to the concession stand, which starts at the ticket booth?

The distance from the ticket booth to the center of the concession stand is given in the problem as 25 feet.

You can draw a radius from the point of tangency of the walkway to the center of the concession stand, which is 7 feet, as it is the same length of the other radius in the circle, which is marked on the diagram.

Use the Pythagorean Theorem to find the length of the walkway.

$a^2 + b^2 = c^2$	Pythagorean Theorem
$7^2 + b^2 = 25^2$	Substitute values for *a* and *c*.
$49 + b^2 = 625$	Simplify, then solve for *b*.
$b^2 = 576$	
$b = 24$	

The length of the walkway tangent to the concession stand is 24 feet.

b. What is the length of the walkway tangent to the concession stand, which ends at the petting zoo?

The radius drawn to the point of tangency of this line is also 7 feet as shown in the diagram and the distance from the center of the concession stand to the petting zoo is 25 feet; therefore, the tangent line is the same length as the one that extends from the ticket booth in part a, 24 feet.

c. What is the length of the walkway tangent to the big top, which extends from the same point as the walkway that is 13 feet long?

Two lines extending from the same exterior point, tangent to the same circle, are congruent.

Since one of the tangents is 13 feet long, the other is also 13 feet long.

d. What is the value for x for the two walkways tangent to the big top, extending from the same point at the ticket booth?

Since two lines extending from the same exterior point to the same circle are congruent, set the equations for the lengths of each of the lines equal to each other to find x.

$2x - 7 = x + 23$

$2x = x + 30$

$x = 30$

e. What is the length of the lines tangent to the big top, extending from the same point at the ticket booth?

Substitute 30 into one of the equations to find the length of the tangent line.

$2x - 7$

$2(30) - 7 = 53$

Although the other tangent is 53 feet as well, it is always good to check by substituting the value of x into the other equation to verify that they are the same.

$x + 23$

$30 + 23 = 53$

The lines tangent to the big top, extending from the same point as the ticket booth, are each 53 feet long.

f. What is the sum of the lengths of all of the exterior walkways?

Add up all of the lengths of all the walkways.

$24 + 24 + 13 + 13 + 53 + 53 = 180$

The sum of the lengths of the walkways is 180 feet.

g. Will the circus pass inspection?

Yes, the circus will pass inspection; the requirement was to have at least 175 feet of exterior walkways, and it has 180 feet of walkways.

G–C.3 • Geometry

Constructing Inscribed Circles

Common Core State Standard

G–C.3

Construct the inscribed and circumscribed circles of a triangle, and prove properties of angles for a quadrilateral inscribed in a circle.

Problem-Based Task: First Aid Station

Organizers are setting up a first aid station and roping off paths for a cross-country race. There are three routes that the runners will travel at least twice throughout the race. Where should the organizers place the first aid station so that help can be provided easily? Where should the roped paths be placed for the responders?

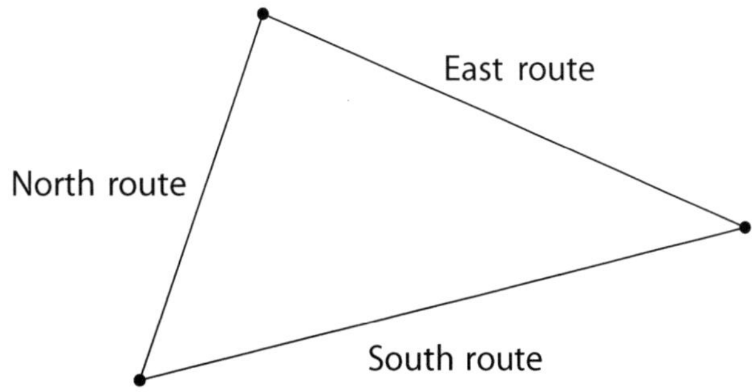

Problem-Based Task: First Aid Station

Coaching

a. How many angles of the triangle need to be bisected in order to find the incenter?

b. Why is only one bisector not sufficient?

c. How many perpendicular segments need to be constructed from the incenter to a side?

d. How can the other point of intersection of the perpendicular segments be found quickly?

e. Where are the roped paths?

Problem-Based Task: First Aid Station

Coaching Sample Responses

a. How many angles of the triangle need to be bisected in order to find the incenter?

Two; since all three bisectors are concurrent, the third is not needed but can be used as a check.

b. Why is only one bisector not sufficient?

One angle bisector shows the ray on which the incenter lies, but a second bisector is needed to find the exact location.

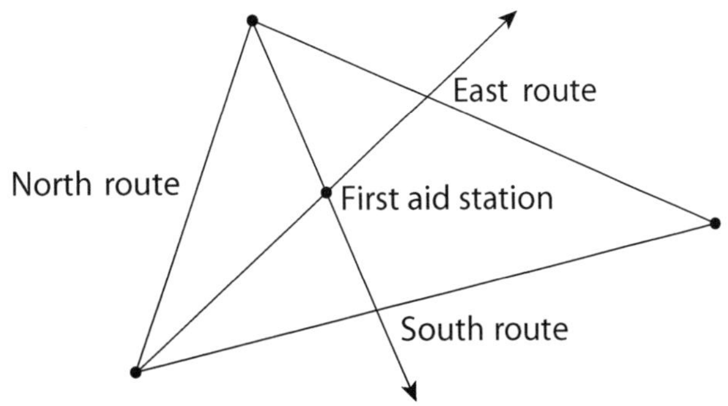

c. How many perpendicular segments need to be constructed from the incenter to a side?

Since the incenter is equidistant from the sides, one perpendicular is sufficient to set the radius of the inscribed circle.

All other distances will be equal.

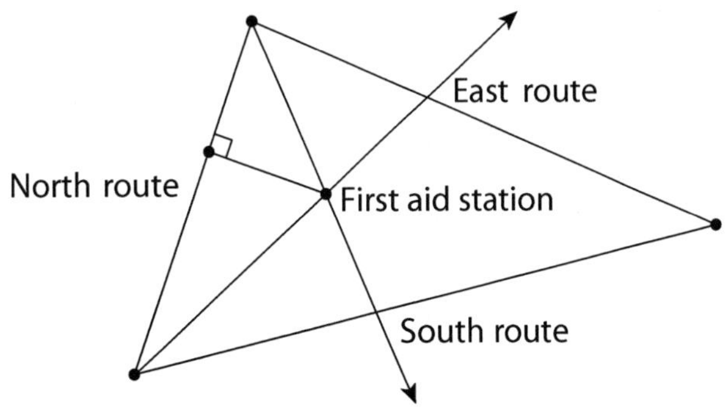

d. How can the other point of intersection of the perpendicular segments be found quickly?

Construct a circle within the triangle; the intersections of the circle with the sides of the triangle will reveal the other points.

e. Where are the roped paths?

The roped paths will be located along the three perpendicular segments.

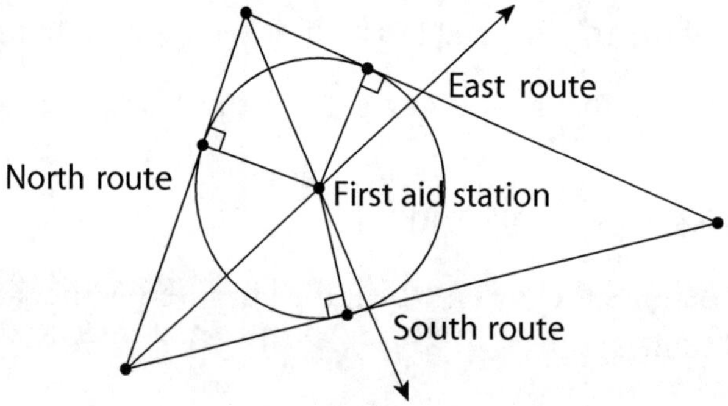

NAME:

G–C.3 • Geometry
Constructing Circumscribed Circles

Common Core State Standard

G–C.3

Construct the inscribed and circumscribed circles of a triangle, and prove properties of angles for a quadrilateral inscribed in a circle.

Problem-Based Task: Building a New Radio Station

The owners of a radio station in Georgia want to build a new broadcasting building located within the triangle formed by Atlanta, Columbus, and Macon. Where should the station be built so that it is equidistant from each city?

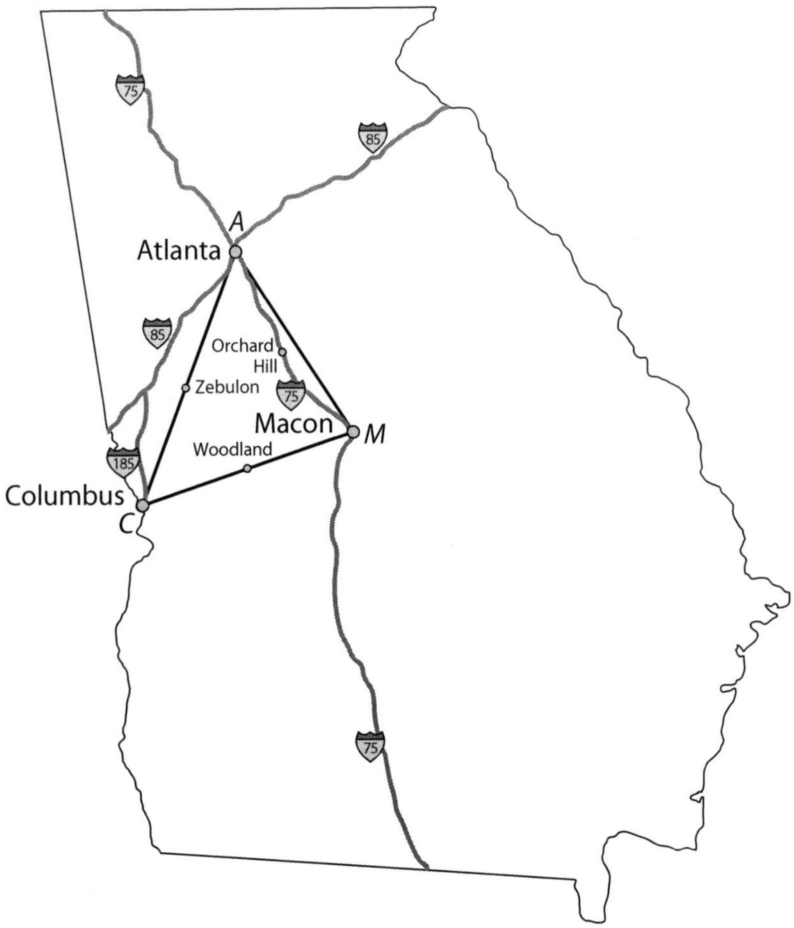

Problem-Based Task: Building a New Radio Station

Coaching

a. Does the new station need to be placed equidistant from the interstates or from the cities?

b. Which is needed to determine the placement of the station: the angle bisectors or the perpendicular bisectors?

c. How many bisectors need to be constructed in order to locate the point of concurrency?

d. Perform the construction. Which town is closest to the point of concurrency?

Problem-Based Task: Building a New Radio Station

Coaching Sample Responses

a. Does the new station need to be placed equidistant from the interstates or from the cities?

The new station needs to be located equidistant from the cities.

b. Which is needed to determine the placement of the station: the angle bisectors or the perpendicular bisectors?

The perpendicular bisectors will locate the point that is equidistant from the cities.

c. How many bisectors need to be constructed in order to locate the point of concurrency?

Only two of the three perpendicular bisectors need to be constructed.

The third perpendicular bisector will also contain the point of intersection determined by the first two perpendicular bisectors.

d. Perform the construction. Which town is closest to the point of concurrency?

Zebulon is closest to the point of concurrency.

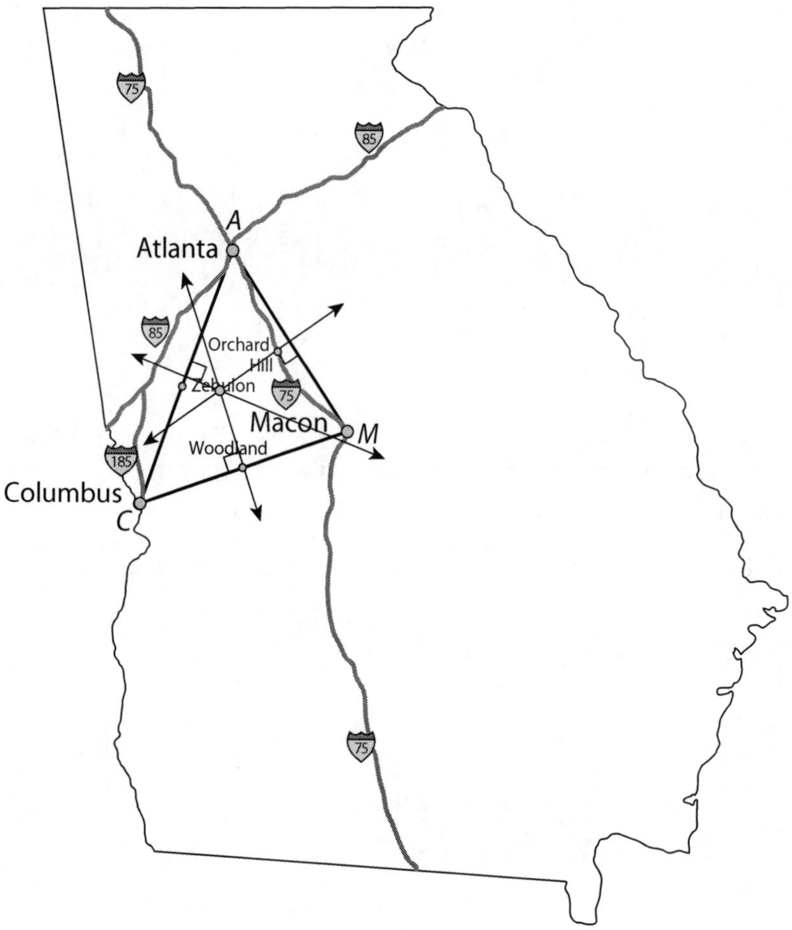

G–C.3 • Geometry

Proving Properties of Inscribed Quadrilaterals

Common Core State Standard

G–C.3

Construct the inscribed and circumscribed circles of a triangle, and prove properties of angles for a quadrilateral inscribed in a circle.

Problem-Based Task: King Arthur and His Round Table

King Arthur is trying to assign seats at his Round Table for three of his knights and himself. If the angle formed by Galahad, Arthur, and Lancelot is 112°, is there enough information to determine the measure of the angle formed by Galahad, Bedivere, and Lancelot? Is there enough information to determine the measures of the angles at Galahad and Lancelot? Also, since Arthur is a geometry fan, he would like to see what types of quadrilaterals he and his knights could form.

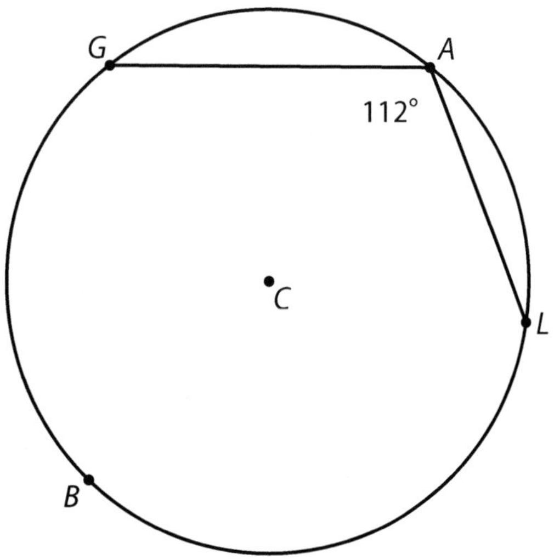

Problem-Based Task: King Arthur and His Round Table

Coaching

a. What is true about the opposite angles of an inscribed quadrilateral?

b. What is the relationship between the angles at *G* and at *L*?

c. Could the knights form a rectangle? If so, how?

d. Could the knights form a square? If so, how?

e. Could the knights form a trapezoid? If so, how?

Problem-Based Task: King Arthur and His Round Table

Coaching Sample Responses

a. What is true about the opposite angles of an inscribed quadrilateral?

The opposite angles of an inscribed quadrilateral are supplementary.

b. What is the relationship between the angles at *G* and at *L*?

The only relationship between those two angles is that they are supplementary.

c. Could the knights form a rectangle? If so, how?

Yes, if all of the angles are 90° but the distances between the pairs of knights are different.

This implies that the diagonals contain the center of the circle.

d. Could the knights form a square? If so, how?

Yes, if all of the angles are 90° and the distances between the knights are the same.

e. Could the knights form a trapezoid? If so, how?

Yes, if the consecutive angles of the inscribed quadrilateral are also supplementary. This would imply an isosceles trapezoid.

NAME:

G–C.4(+) • Geometry

Constructing Tangent Lines

Common Core State Standard

G–C.4

(+) Construct a tangent line from a point outside a given circle to the circle.

Problem-Based Task: Designing a T-Shirt Logo

A local T-shirt shop is designing a logo for a company. The company's logo is a circle inscribed in a triangle. The sketch below is incomplete. The designer needs to draw lines tangent to the circle from point *C* to point *A* and from point *C* to point *B*.

The designer also needs to calculate the perimeter of the triangle to be sure the logo fits in the allotted space. He knows the following information about the dimensions of the design:

- The diameter of circle *G* is 8 cm.
- $\overline{CI} = 4$ cm
- $\overline{CI}$, $\overline{BH}$, and $\overline{AJ}$ are all congruent.

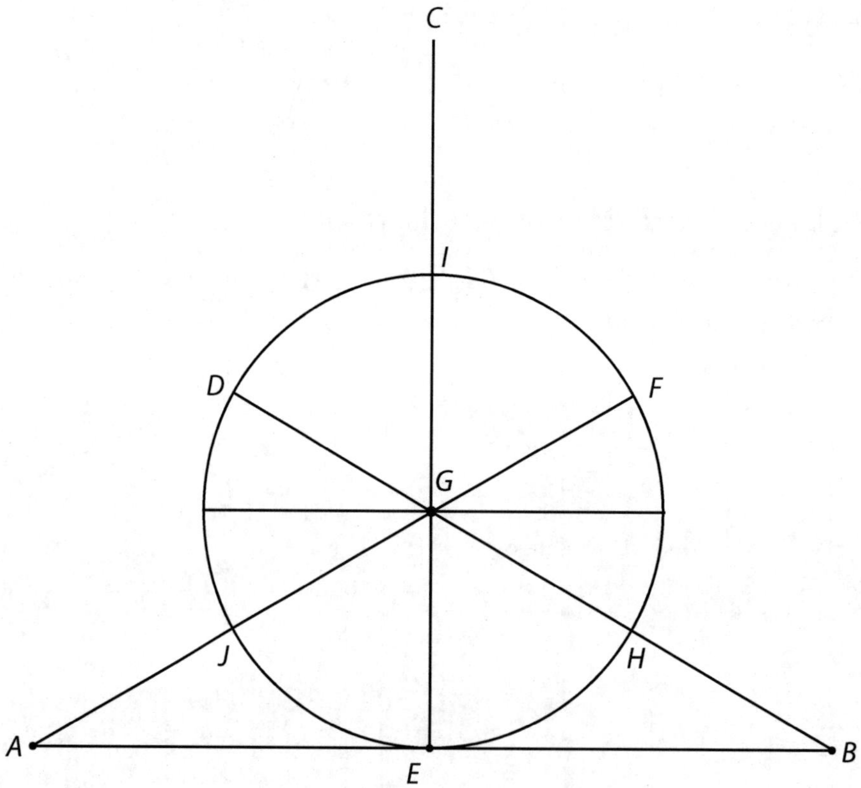

Problem-Based Task: Designing a T-Shirt Logo

Coaching

a. How can the designer draw the tangents $\overline{CA}$ and $\overline{CB}$ so the design can be completed?

b. What is the length of the radius of the circle?

c. What is the length of $\overline{GC}$?

d. How do you calculate the length of $\overline{CF}$?

e. How does the length of $\overline{CF}$ help you to know the length of $\overline{CD}$?

f. If $\overline{CI}$, $\overline{BH}$, and $\overline{AJ}$ are all congruent, what are the lengths of $\overline{BF}$ and $\overline{AD}$?

g. What are the lengths of $\overline{AE}$ and $\overline{BE}$?

h. What is the perimeter of $\triangle ABC$?

Problem-Based Task: Designing a T-Shirt Logo

Coaching Sample Responses

a. How can the designer draw the tangents $\overline{CA}$ and $\overline{CB}$ so the design can be completed?

Find the midpoint of $\overline{GC}$ by constructing the perpendicular bisector of the line.

Place the sharp edge of the compass on the midpoint and set the length of the compass to center *G* of the circle.

Without changing the compass width, draw an arc on the circle in two possible places. These are the points of tangency.

Draw a tangent line from *C* through points *D* and *F*, and from *C* through points *A* and *B*.

b. What is the length of the radius of the circle?

4 cm (half of the diameter)

c. What is the length of $\overline{GC}$?

$4 + 4 = 8$

d. How do you calculate the length of $\overline{CF}$?

Use the Pythagorean Theorem to calculate the length of $\overline{CF}$.

$a^2 + b^2 = c^2$	Pythagorean Theorem
$4^2 + (CF)^2 = 8^2$	Substitute.
$16 + (CF)^2 = 64$	Simplify.
$(CF)^2 = 48$	

$CF = \sqrt{48} = 4\sqrt{3} \approx 6.9$

The length of *CF* is $4\sqrt{3}$ or approximately 6.9 cm.

e. How does the length of $\overline{CF}$ help you to know the length of $\overline{CD}$?

Since $\overline{CF}$ and $\overline{CD}$ extend from the same exterior point and are tangent to the same circle, the line segments are congruent. $CF = CD$; therefore, $CD = 4\sqrt{3}$ or approximately 6.9 cm.

f. If $\overline{CI}$, $\overline{BH}$, and $\overline{AJ}$ are all congruent, what are the lengths of $\overline{BF}$ and $\overline{AD}$?

The radii are all the same length regardless of the point on the circle to which they extend. $\overline{CI}$, $\overline{BH}$, and $\overline{AJ}$ are all 4 cm, so the lengths of $\overline{BF}$ and $\overline{AD}$ would be calculated exactly as the length of $\overline{CF}$ was calculated in part d. Therefore, $\overline{BF}$ and $\overline{AD}$ are both $4\sqrt{3}$ or approximately 6.9 cm.

g. What are the lengths of $\overline{AE}$ and $\overline{BE}$?

$\overline{AE}$ and $\overline{BE}$ are both $4\sqrt{3}$ or approximately 6.9 cm.

h. What is the perimeter of $\triangle ABC$?

To find the perimeter of the triangle, find the sum of the segments.

$$AE + AD + BE + BF + CD + CF = 4\sqrt{3} + 4\sqrt{3} + 4\sqrt{3} + 4\sqrt{3} + 4\sqrt{3} + 4\sqrt{3} = 24\sqrt{3} \approx 41.6$$

The perimeter of the triangle is $24\sqrt{3}$ or approximately 41.6 cm.

G–C.5 • Geometry
Defining Radians

Common Core State Standard

G–C.5

Derive using similarity the fact that the length of the arc intercepted by an angle is proportional to the radius, and define the radian measure of the angle as the constant of proportionality; derive the formula for the area of a sector.

Problem-Based Task: Around the Merry-Go-Round

A merry-go-round has a circumference of 160 feet and revolves at a speed of 6 miles per hour. How many feet does a carousel horse on the outer edge of the merry-go-round travel in a 2-minute ride? How many radians does a carousel horse travel in a 2-minute ride?

Problem-Based Task: Around the Merry-Go-Round

Coaching

a. How many miles does the carousel horse travel in 1 minute?

b. How many feet does the carousel horse travel in 1 minute?

c. How many feet does the carousel horse travel in 2 minutes?

d. What is the radius of the merry-go-round?

e. Use the distance traveled and the radius of the merry-go-round to find the radian measure of the angle.

f. How many revolutions does the merry-go-round make in this time? Does your answer seem reasonable?

Problem-Based Task: Around the Merry-Go-Round

Coaching Sample Responses

a. How many miles does the carousel horse travel in 1 minute?

Divide the speed by 60 minutes to convert from miles per hour to miles per minute.

$$\frac{6 \text{ miles}}{1 \cancel{\text{hour}}} \bullet \frac{1 \cancel{\text{hour}}}{60 \text{ minutes}} = \frac{1}{10} \text{ mile per minute}$$

b. How many feet does the carousel horse travel in 1 minute?

There are 5,280 feet in 1 mile. Multiply the speed by 5,280 feet to convert from miles per minute to feet per minute.

$$\frac{1 \cancel{\text{mile}}}{10 \text{ minutes}} \bullet \frac{5280 \text{ feet}}{1 \cancel{\text{mile}}} = 528 \text{ feet per minute}$$

c. How many feet does the carousel horse travel in 2 minutes?

Multiply the number of feet traveled in 1 minute by 2 to find the number of feet traveled in 2 minutes.

$528 \bullet 2 = 1056$ feet

d. What is the radius of the merry-go-round?

Find the circumference.

$C = 160 \text{ feet} = 2\pi r$

Divide each side by 2π to find the radius.

$$r = \frac{160}{2\pi} = \frac{80}{\pi} \approx 25.46 \text{ feet}$$

e. Use the distance traveled and the radius of the merry-go-round to find the radian measure of the angle.

Divide the distance traveled (the arc length) by the radius to find the radian measure of the angle. Use $\frac{80}{\pi}$ for the radius since this value is more accurate.

$$\theta = \frac{s}{r} = \frac{1056}{\frac{80}{\pi}} = \frac{66\pi}{5} \text{ radians} \approx 41.47 \text{ radians}$$

f. How many revolutions does the merry-go-round make in this time? Does your answer seem reasonable?

To find the number of revolutions in 2 minutes, divide the radian measure by 2π, which is one revolution. Use $\frac{66\pi}{5}$ since this value is more accurate.

$$\frac{\frac{66\pi}{5}}{2\pi} \approx 6$$

The merry-go-round goes through approximately 6.6 revolutions in 2 minutes. This seems reasonable.

G–C.5 • Geometry

Deriving the Formula for the Area of a Sector

Common Core State Standard

G–C.5

Derive using similarity the fact that the length of the arc intercepted by an angle is proportional to the radius, and define the radian measure of the angle as the constant of proportionality; derive the formula for the area of a sector.

Problem-Based Task: Pizza Specials

A pizza parlor has a \$5 lunch special this week for 3 pieces of a small cheese pizza and a soft drink. Next week, the \$5 special will be for 2 pieces of a large cheese pizza and a soft drink. A small pizza measures 10 inches in diameter and is cut into 6 equal slices. A large pizza measures 14 inches in diameter and is cut into 8 equal slices. Which special is the better deal?

Problem-Based Task: Pizza Specials

Coaching

a. What is the radius of a small pizza?

b. What is the area of a small pizza?

c. One slice of a small pizza makes up what fraction of a whole small pizza?

d. What is the area of a 1-piece sector of a small pizza?

e. What is the area of 3 pieces of a small pizza?

f. What is the radius of a large pizza?

g. What is the area of a large pizza?

h. One slice of a large pizza makes up what fraction of a whole large pizza?

i. What is the area of a 1-piece sector of a large pizza?

j. What is the area of 2 pieces of a large pizza?

k. How does the area of 3 pieces of a small pizza compare to the area of 2 pieces of a large pizza?

l. What can you conclude about the lunch specials?

Problem-Based Task: Pizza Specials

Coaching Sample Responses

a. What is the radius of a small pizza?

$$r = \frac{1}{2} \bullet d = \frac{1}{2} \bullet 10 = 5 \text{ inches}$$

b. What is the area of a small pizza?

$$A_{\text{small}} = \pi r^2 = \pi \bullet 5^2 = 25\pi \approx 78.54 \text{ square inches}$$

c. One slice of a small pizza makes up what fraction of a whole small pizza?

A small pizza is cut into 6 equal slices, so 1 slice is $\frac{1}{6}$ of a whole pizza.

d. What is the area of a 1-piece sector of a small pizza?

$$\frac{1}{6} \bullet A_{\text{small}} = \frac{1}{6} \bullet 25\pi = \frac{25\pi}{6} \approx 13.09 \text{ square inches}$$

e. What is the area of 3 pieces of a small pizza?

$$3 \bullet \frac{25\pi}{6} = \frac{25\pi}{2} \approx 39.27 \text{ square inches}$$

f. What is the radius of a large pizza?

$$r = \frac{1}{2} \bullet d = \frac{1}{2} \bullet 14 = 7 \text{ inches}$$

g. What is the area of a large pizza?

$$A_{\text{large}} = \pi r^2 = \pi \bullet 7^2 = 49\pi \approx 153.94 \text{ square inches}$$

h. One slice of a large pizza makes up what fraction of a whole large pizza?

A large pizza is cut into 8 equal slices, so 1 slice is $\frac{1}{8}$ of a whole pizza.

i. What is the area of a 1-piece sector of a large pizza?

$$\frac{1}{8} \bullet A_{\text{large}} = \frac{1}{8} \bullet 49\pi = \frac{49\pi}{8} \approx 19.24 \text{ square inches}$$

j. What is the area of 2 pieces of a large pizza?

$$2 \bullet \frac{49\pi}{8} = \frac{49\pi}{4} \approx 38.48 \text{ square inches}$$

k. How does the area of 3 pieces of a small pizza compare to the area of 2 pieces of a large pizza?

The area of 3 pieces of a small pizza is about 39.27 square inches. This is greater than the area of 2 pieces of a large pizza, which is about 38.48 square inches.

l. What can you conclude about the lunch specials?

The lunch special this week is a slightly better deal than the lunch special next week because the area of 3 pieces of a small pizza is slightly greater than the area of 2 pieces of a large pizza.

G–GPE.1 • Geometry

Deriving the Equation of a Circle

Common Core State Standard

G–GPE.1

Derive the equation of a circle of given center and radius using the Pythagorean Theorem; complete the square to find the center and radius of a circle given by an equation.

Problem-Based Task: Nurturing an Investment

Anna's landscaping company has a contract to improve and maintain a municipal park. Anna made a scale drawing of the park on a coordinate system, using meters as the unit of distance. She has already installed two permanent sprinkler outlets. Sprinkler 1 waters inside the region whose boundary has the equation $x^2 + y^2 - 20x - 20y + 136 = 0$. Sprinkler 2 waters inside the region whose boundary has the equation $x^2 + y^2 - 50x - 24y + 669 = 0$. Anna bought an expensive tree and she wants to plant it at the point (17, 8), where she thinks it will be watered by both sprinklers. Will the tree be watered by both sprinklers at that point? Draw a sketch that illustrates your answer.

Problem-Based Task: Nurturing an Investment

Coaching

a. What geometric figures are represented by the given equations? How do you know?

b. What facts do you need to know in order to determine the regions watered by the sprinklers?

c. What form of equation do you need to find the facts referenced in part b?

d. What are the main steps needed to obtain the form referenced in part c?

e. For sprinkler 1, apply the necessary steps to obtain the form referenced in parts c and d.

f. What is the geometric description of the region watered by sprinkler 1?

g. What is the geometric description of the region watered by sprinkler 2?

h. What is the distance between sprinkler 1 and the planned location of the tree?

i. Will the tree be watered by sprinkler 1? Explain.

j. Will the tree be watered by sprinkler 2? Explain.

k. Will the tree be watered by both sprinklers? Draw a sketch that illustrates your answer.

Problem-Based Task: Nurturing an Investment

Coaching Sample Responses

a. What geometric figures are represented by the given equations? How do you know?

The figures are circles. The equations are in the general form of an equation of a circle.

b. What facts do you need to know in order to determine the regions watered by the sprinklers?

You need to know the center and radius of each circle.

c. What form of equation do you need to find the facts referenced in part b?

You need the standard form of an equation of a circle.

d. What are the main steps needed to obtain the form referenced in part c?

For each given equation, add or subtract as needed from both sides to get the constant term on one side. Group same-variable terms. Complete the square for both variables, making sure to add the same values to both sides of the equation. Write the perfect square trinomials as squares of binomials.

e. For sprinkler 1, apply the necessary steps to obtain the form referenced in parts c and d.

$$x^2 + y^2 - 20x - 20y + 136 = 0$$

$$x^2 + y^2 - 20x - 20y = -136$$

$$x^2 - 20x + y^2 - 20y = -136$$

Next, complete the square for both variables. Add the same values to both sides of the equation, as shown below:

$$x^2 - 20x + \left(\frac{-20}{2}\right)^2 + y^2 - 20y + \left(\frac{-20}{2}\right)^2 = -136 + \left(\frac{-20}{2}\right)^2 + \left(\frac{-20}{2}\right)^2$$

$$x^2 - 20x + 100 + y^2 - 20y + 100 = -136 + 100 + 100$$

$$(x - 10)^2 + (y - 10)^2 = 64$$

The standard equation is $(x - 10)^2 + (y - 10)^2 = 64$.

f. What is the geometric description of the region watered by sprinkler 1?

Use the standard equation to determine the center and radius of the region watered by sprinkler 1. Sprinkler 1 waters the interior of the circle with center (10, 10) and radius 8 meters.

g. What is the geometric description of the region watered by sprinkler 2?

Convert the general equation to standard form, and then determine the center and radius of the region watered by sprinkler 2.

$$x^2 + y^2 - 50x - 24y + 669 = 0$$

$$x^2 + y^2 - 50x - 24y = -669$$

$$x^2 - 50x + y^2 - 24y = -669$$

Next, complete the square for both variables. Add the same values to both sides of the equation, as shown:

$$x^2 - 50x + \left(\frac{-50}{2}\right)^2 + y^2 - 24y + \left(\frac{-24}{2}\right)^2 = -669 + \left(\frac{-50}{2}\right)^2 + \left(\frac{-24}{2}\right)^2$$

$$x^2 - 50x + 625 + y^2 - 24y + 144 = -669 + 625 + 144$$

$$(x - 25)^2 + (y - 12)^2 = 100$$

Sprinkler 2 waters the interior of the circle with center (25, 12) and radius 10 meters.

h. What is the distance between sprinkler 1 and the planned location of the tree?

Use the distance formula, substituting points (10, 10) and (17, 8) for (x_1, y_1) and (x_2, y_2).

$$d = \sqrt{(17-10)^2 + (8-10)^2}$$

$$d = \sqrt{49+4}$$

$$d = \sqrt{53} \text{ meters}$$

i. Will the tree be watered by sprinkler 1? Explain.

Yes; the radius of the circular region watered by sprinkler 1 is 8 meters, or $\sqrt{64}$ meters. The tree will be $\sqrt{53}$ meters from the sprinkler, which is less than the radius.

j. Will the tree be watered by sprinkler 2? Explain.

Use the distance formula, substituting points (25, 12) and (17, 8) for (x_1, y_1) and (x_2, y_2).

$$d=\sqrt{(17-25)^2+(8-12)^2}$$

$$d=\sqrt{64+16}$$

$$d=\sqrt{80}\text{ meters}$$

Yes; the radius of the circular region watered by sprinkler 2 is 10 meters, or $\sqrt{100}$ meters. The tree will be $\sqrt{80}$ meters from the sprinkler, which is less than the radius.

k. Will the tree be watered by both sprinklers? Draw a sketch that illustrates your answer.

Yes, the tree will be watered by both sprinklers. Sample sketch:

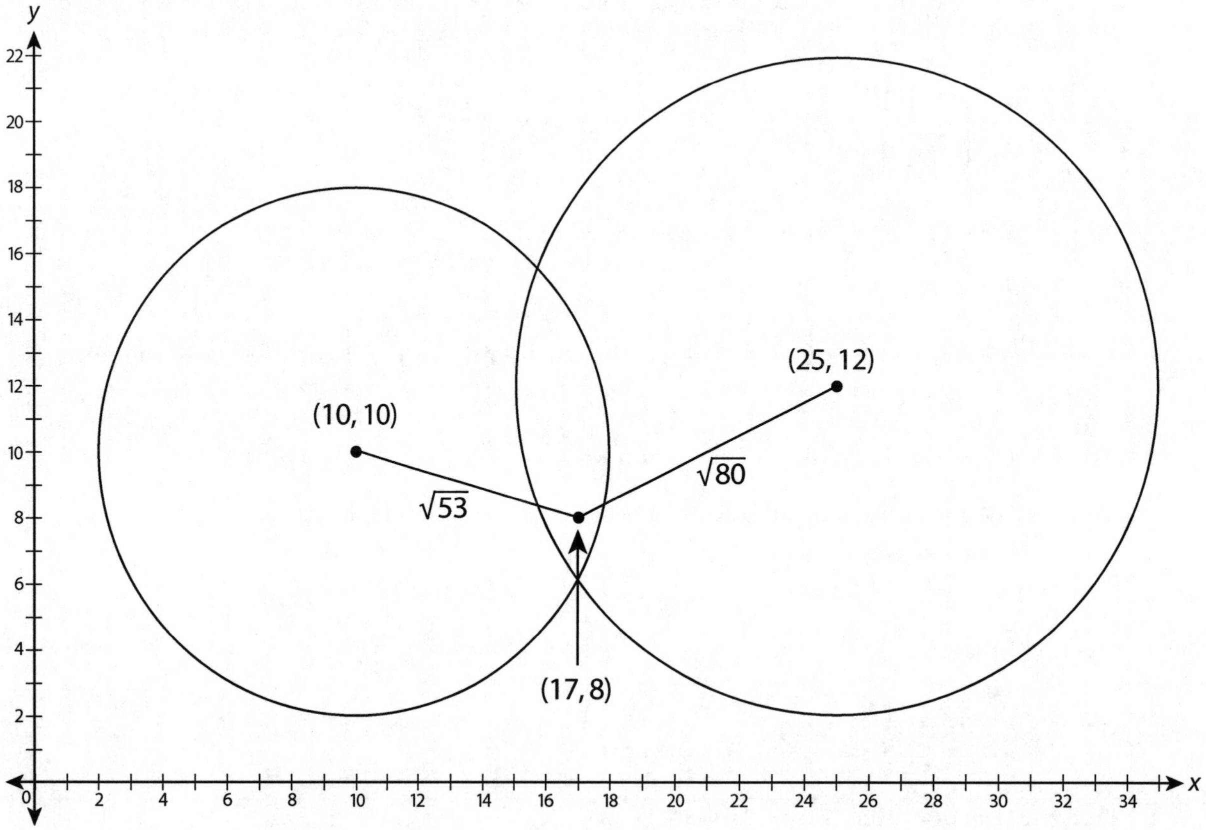

Common Core State Standard

G–GPE.2

Derive the equation of a parabola given a focus and directrix.

Problem-Based Task: A Ball in Flight

A softball player hits the ball when it is directly above home plate. The ball then follows a parabolic path until it hits the ground. The path of the ball is represented by a portion of a parabola on a coordinate plane, where the origin represents home plate and the unit of distance is feet. The focus of the parabola is (100, –18.5) and the directrix is $y = 106.5$.

Write the equation of the path of the ball and supply the following key facts: the maximum height of the ball, the height of the ball when it is hit by the bat, and how far the ball travels horizontally before it hits the ground.

Problem-Based Task: A Ball in Flight

Coaching

a. Make a sketch using the given information.

b. What is the *x*-coordinate of the vertex? Explain.

c. In what direction does the parabola open? Explain.

d. What is the value of *p* for this parabola? Explain.

e. What is the *y*-coordinate of the vertex? Explain.

f. Summarize what you have found.

g. What is the equation of the parabola? Show your work.

h. What is the maximum height of the ball? Explain.

i. Where on the field is the ball when it is hit by the bat?

j. What point on the graph represents that location? Explain.

k. What is the height of the ball when it is hit by the bat? Explain.

l. What point on the graph represents the point where the ball hits the ground?

m. How can you find the coordinates of the points where the graph intersects the *x*-axis?

n. What are the approximate coordinates of those points? Show your work.

o. Sketch the graph of the parabola. Plot and label all the key points you have found. Identify the portion of the graph that represents the path of the ball.

p. How far does the ball travel horizontally before it hits the ground? Explain.

Problem-Based Task: A Ball in Flight

Coaching Sample Responses

a. Make a sketch using the given information.

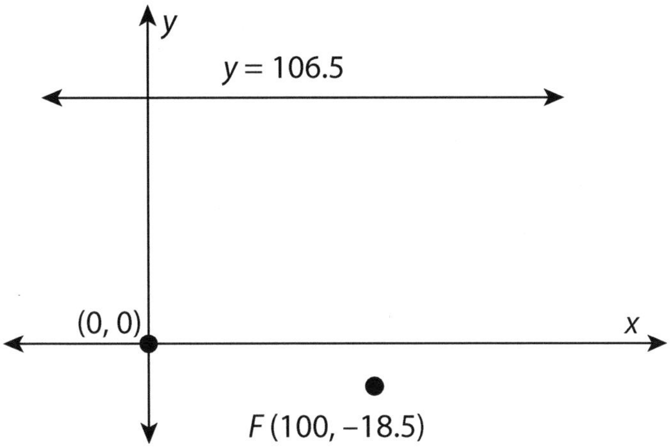

b. What is the x-coordinate of the vertex? Explain.

The x-coordinate is 100 because the vertex is directly above the focus.

c. In what direction does the parabola open? Explain.

The parabola opens down because the focus is below the directrix.

d. What is the value of p for this parabola? Explain.

$p = -62.5$. The distance between the focus and the directrix is $2|p|$, so $2|p| = 106.5 - (-18.5) = 125$, and $|p| = 62.5$. The parabola opens down, so p is negative.

e. What is the y-coordinate of the vertex? Explain.

The y-coordinate is 44. The distance between the focus and the vertex is $|p|$, which is 62.5 in this case. So the vertex is 62.5 units above the focus. Add to find the y-coordinate:

$-18.5 + 62.5 = 44$

f. Summarize what you have found.

The vertex is (100, 44) and $p = -62.5$.

g. What is the equation of the parabola? Show your work.

The equation is $(x - 100)^2 = -250(y - 44)$. This can be determined as follows:

$$(x - h)^2 = 4p(y - k)$$

$$(x - 100)^2 = 4(-62.5)(y - 44)$$

$$(x - 100)^2 = -250(y - 44)$$

h. What is the maximum height of the ball? Explain.

The maximum height is 44 feet because the vertex is (100, 44).

i. Where on the field is the ball when it is hit by the bat?

It is directly above home plate.

j. What point on the graph represents that location? Explain.

The point where the graph intersects the y-axis represents this location. The origin represents home plate and the ball is directly above home plate when it is hit, so the corresponding point on the graph is on the y-axis.

k. What is the height of the ball when it is hit by the bat? Explain.

The ball is 4 feet above the ground when it is hit. This is determined by finding the y-intercept, so substitute 0 for x in the equation.

$$(x - 100)^2 = -250(y - 44)$$

$$(0 - 100)^2 = -250(y - 44)$$

$$10{,}000 = -250(y - 44)$$

$$\frac{10{,}000}{-250} = \frac{-250}{-250}(y - 44)$$

$$-40 = y - 44$$

$$4 = y$$

l. What point on the graph represents the point where the ball hits the ground?

One of the points where the graph intersects the x-axis.

m. How can you find the coordinates of the points where the graph intersects the *x*-axis?

Substitute 0 for *y* in the equation.

n. What are the approximate coordinates of those points? Show your work.

(–5, 0) and (205, 0). This can be determined as follows:

$(x-100)^2 = -250(y-44)$

$(x-100)^2 = -250(0-44)$

$(x-100)^2 = 11{,}000$

$x-100 = \pm\sqrt{11{,}000}$

$x = 100 \pm \sqrt{11{,}000}$

$x = 100 + \sqrt{11{,}000} \approx 205$ or $x = 100 - \sqrt{11{,}000} \approx -5$

o. Sketch the graph of the parabola. Plot and label all the key points you have found. Identify the portion of the graph that represents the path of the ball.

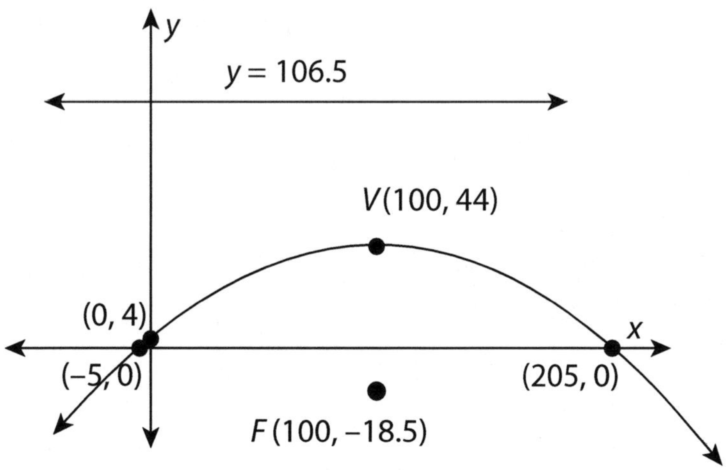

The path of the ball is represented by the portion of the parabola from (0, 4), where it is hit by the bat, to (205, 0), where it hits the ground.

p. How far does the ball travel horizontally before it hits the ground? Explain.

The ball travels approximately 205 feet horizontally because that is the horizontal distance between (0, 4) and (205, 0).

G–GPE.4 • Geometry

Using Coordinates to Prove Geometric Theorems About Circles and Parabolas

Common Core State Standard

G–GPE.4

Use coordinates to prove simple geometric theorems algebraically. *For example, prove or disprove that a figure defined by four given points in the coordinate plane is a rectangle; prove or disprove that the point* $(1, \sqrt{3})$ *lies on the circle centered at the origin and containing the point (0, 2).*

Problem-Based Task: A Circle Graph for Lunch

Alena conducted a survey of her classmates' lunch preferences. She drew a circle graph on a transparency sheet to show the results. She placed the circle graph on a coordinate plane and positioned it so that points *R* and *T* have the coordinates *R* (24, 22) and *T* (8, 10). Alena claims that the coordinates of point *P* are (8, 22). Is she correct? Prove or disprove her claim.

Lunch Preferences

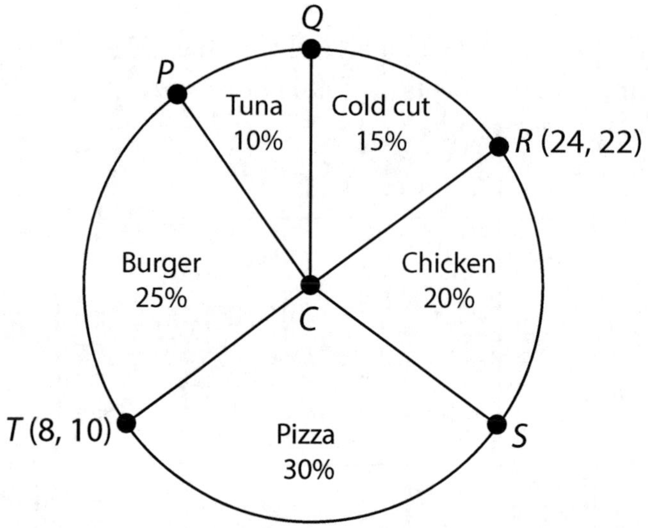

Problem-Based Task: A Circle Graph for Lunch

Coaching

a. What are the two key facts needed for a complete geometric description of this circle on a coordinate plane?

b. What are the only coordinates you are given?

c. What fact about line segment *RT* will be useful in determining the coordinates of *C*, the center of the circle?

d. How do you know this fact about $\overline{RT}$?

e. How can you find the coordinates of *C*?

f. What are the coordinates of *C*? Explain.

g. What is the radius of the circle? Explain.

h. Now you know the radius and center of the circle. What condition must the point (8, 22) satisfy to be on the circle?

i. Does the point (8, 22) satisfy that condition? Explain.

j. Have you proved that *P* is at (8, 22)? Explain.

k. How can you prove that (8, 22) is the endpoint of $\overline{PC}$, the radius segment that separates the "burger" sector and the "tuna" sector? Explain.

l. Is Alena's claim correct? Explain.

Problem-Based Task: A Circle Graph for Lunch

Coaching Sample Responses

a. What are the two key facts needed for a complete geometric description of this circle on a coordinate plane?

The radius and the coordinates of the center are needed.

b. What are the only coordinates you are given?

You are given *R* (24, 22) and *T* (8, 10).

c. What fact about line segment *RT* will be useful in determining the coordinates of *C*, the center of the circle?

$\overline{RT}$ is the diameter of the circle.

d. How do you know this fact about $\overline{RT}$?

The "chicken" and "pizza" sectors of the circle amount to 50%.

e. How can you find the coordinates of *C*?

Find the coordinates of the midpoint of $\overline{RT}$.

f. What are the coordinates of *C*? Explain.

C is the midpoint of $\overline{RT}$, so use the midpoint formula to find its coordinates.

R is at (24, 22) and *T* is at (8, 10).

$$\left(\frac{x_1+x_2}{2},\frac{y_1+y_2}{2}\right)$$

$$\left(\frac{8+24}{2},\frac{10+22}{2}\right)=(16,16)$$

C is at (16, 16).

g. What is the radius of the circle? Explain.

The radius is the distance from *T* to the center *C*, which is found using the distance formula:

$$CT=\sqrt{(16-8)^2+(16-10)^2}=\sqrt{64+36}=\sqrt{100}=10$$

The radius is 10.

h. Now you know the radius and center of the circle. What condition must the point (8, 22) satisfy to be on the circle?

It must be 10 units from the center, (16, 16).

i. Does the point (8, 22) satisfy that condition? Explain.

Yes; using the distance formula, we can determine that (8, 22) is 10 units from (16, 16).

$$\sqrt{(16-8)^2+(16-22)^2}=\sqrt{64+36}=\sqrt{100}=10$$

j. Have you proved that *P* is at (8, 22)? Explain.

No; we have proved that (8, 22) is on the circle, but not that *P* is at (8, 22). It is necessary to prove that (8, 22) is the endpoint of the radius segment that separates the "burger" sector and the "tuna" sector.

k. How can you prove that (8, 22) is the endpoint *P* of $\overline{PC}$, the radius segment that separates the "burger" sector and the "tuna" sector? Explain.

The "burger" sector is 25%, so $\overline{PC}$ is perpendicular to $\overline{CT}$. It is necessary to show that the segment connecting the point (8, 22) and the point *C* (16, 16) is perpendicular to $\overline{CT}$.

l. Is Alena's claim correct? Explain.

No, Alena's claim is not correct.

The slope of the segment connecting (8, 22) and *C* (16, 16) is $\dfrac{16-22}{16-8}=\dfrac{-6}{8}=\dfrac{-3}{4}$.

The slope of the radius segment $\overline{CT}$ is $\dfrac{16-10}{16-8}=\dfrac{6}{8}=\dfrac{3}{4}$.

The segment connecting (8, 22) and *C* (16, 16) is not perpendicular to the radius segment $\overline{CT}$ because $\left(\dfrac{-3}{4}\right)\left(\dfrac{3}{4}\right)\neq -1$.

Therefore, point *P* is not at (8, 22).

G–GPE.6 • Geometry

Midpoints and Other Points on Line Segments

Common Core State Standard

G–GPE.6

Find the point on a directed line segment between two given points that partitions the segment in a given ratio.

Problem-Based Task: Points of Shade

Officials at Serengeti National Park created a map to represent an area of the park. Lion prides are located at the points (–4, –1), (1, 5), (6, 2), and (3, –7). Water holes are located at the points (1, –1) and (8, –4). Acacia trees provide shade for the prides. One tree is located halfway between each of the water holes. What is the location of the acacia tree? How far is this tree from the pride located at (6, 2)? How far is this tree from the pride located at (3, –7)? Describe the location of the acacia tree in terms of the prides located at (6, 2) and (3, –7). Explain your reasoning. Each unit represents 1,000 feet.

Problem-Based Task: Points of Shade

Coaching

a. What are the coordinates of each of the water holes?

b. "Halfway" is another way of describing which mathematical term?

c. Which formula determines the halfway point between two coordinates?

d. What is the halfway point between the two water holes?

e. Which formula determines the distance between two points?

f. What is the distance between the acacia tree and the pride located at (6, 2)?

g. What is the distance between the acacia tree and the pride located at (3, –7)?

h. What do you notice about the location of the acacia tree in terms of the prides located at (6, 2) and (3, –7)?

Problem-Based Task: Points of Shade

Coaching Sample Responses

a. What are the coordinates of each of the water holes?

The coordinates of the water holes are (1, –1) and (8, –4).

b. "Halfway" is another way of describing which mathematical term?

Halfway describes the midpoint.

c. Which formula determines the halfway point between two coordinates?

The midpoint formula determines the halfway point between two coordinates. The formula is $\left(\frac{x_1+x_2}{2},\frac{y_1+y_2}{2}\right)$.

d. What is the halfway point between the two water holes?

Let (x_1, y_1) be (1, –1) and (x_2, y_2) be (8, –4).

$\left(\frac{x_1+x_2}{2},\frac{y_1+y_2}{2}\right)$	Midpoint formula
$\left(\frac{(1)+(8)}{2},\frac{(-1)+(-4)}{2}\right)$	Substitute (1, –1) and (8, –4).
$\left(\frac{9}{2},\frac{-5}{2}\right)$	Simplify.

The midpoint between the two water holes is $\left(\frac{9}{2},\frac{-5}{2}\right)$.

e. Which formula determines the distance between two points?

The distance formula determines the distance between two points. The formula is $\sqrt{(x_2-x_1)^2+(y_2-y_1)^2}$.

f. What is the distance between the acacia tree and the pride located at (6, 2)?

$\sqrt{(x_2-x_1)^2+(y_2-y_1)^2}$ Distance formula

$\sqrt{\left((6)-\left(\frac{9}{2}\right)\right)^2+\left((2)-\left(-\frac{5}{2}\right)\right)^2}$ Substitute $\left(\frac{9}{2},\frac{-5}{2}\right)$ and (6, 2).

$\sqrt{\left(\frac{3}{2}\right)^2+\left(\frac{9}{2}\right)^2}$ Simplify as needed.

$\sqrt{\frac{9}{4}+\frac{81}{4}}$

$\sqrt{\frac{90}{4}}$

$\frac{3\sqrt{10}}{2}$

The distance between the acacia tree and the pride located at (6, 2) is $\frac{3\sqrt{10}}{2}$ units, or approximately 4,743.4 feet.

g. What is the distance between the acacia tree and the pride located at (3, –7)?

$\sqrt{(x_2-x_1)^2+(y_2-y_1)^2}$ Distance formula

$\sqrt{\left((3)-\left(\frac{9}{2}\right)\right)^2+\left((-7)-\left(-\frac{5}{2}\right)\right)^2}$ Substitute $\left(\frac{9}{2},\frac{-5}{2}\right)$ and (3, –7).

$\sqrt{\left(-\frac{3}{2}\right)^2+\left(-\frac{9}{2}\right)^2}$ Simplify as needed.

$\sqrt{\frac{9}{4}+\frac{81}{4}}$

$\sqrt{\frac{90}{4}}$

$\frac{3\sqrt{10}}{2}$

The distance between the acacia tree and the pride located at (3, –7) is $\frac{3\sqrt{10}}{2}$ units, or approximately 4,743.4 feet.

h. What do you notice about the location of the acacia tree in terms of the prides located at (6, 2) and (3, –7)?

The distance from the acacia tree to each of the prides is equal; therefore, the acacia tree is the midpoint between the prides.

G–GMD.1 • Geometry

Circumference and Area of a Circle

Common Core State Standard

G–GMD.1

Give an informal argument for the formulas for the circumference of a circle, area of a circle, volume of a cylinder, pyramid, and cone. *Use dissection arguments, Cavalieri's principle, and informal limit arguments.*

Problem-Based Task: Designing a Tablecloth

A factory is printing a large square design on a circular tablecloth. The square design should be as large as possible. The tablecloth has an area of 30 square feet. What should be the maximum side length of the square design? What area of the tablecloth will not be printed with the square design?

Problem-Based Task: Designing a Tablecloth

Coaching

a. Make a drawing of this problem.

b. What length of the circle corresponds to a length in the square?

c. How can you use the area of the circle to find the diagonal of the square?

d. What angle does the square's diagonal make at the vertex of the side length?

e. What ratio of trigonometry can you use to find the maximum length of the square's side?

f. What is the area of the square?

g. What is the area of the circle minus the area of the square?

Problem-Based Task: Designing a Tablecloth

Coaching Sample Responses

a. Make a drawing of this problem.

b. What length of the circle corresponds to a length in the square?

The diameter of the circle is equal to the diagonal of the square.

c. How can you use the area of the circle to find the diagonal of the square?

The diameter is twice the length of the radius of the tablecloth. The area of the tablecloth is given. Substitute the area of the tablecloth into the formula for the area of a circle.

$A = \pi r^2$

$30 = \pi r^2$

$\frac{30}{\pi} = r^2$

$r = \sqrt{9.549296586}$

$r = \pm 3.090193616$

$r \approx 3.0902$ feet

The diameter is twice the radius, so multiply the radius by 2.

$2 \bullet 3.0902 = 6.1804$ feet

The length of the diagonal of the square design is approximately 6.1804 feet.

d. What angle does the square's diagonal make at the vertex of the side length?

The diagonal bisects the 90° angle in the square. Therefore, each angle measures 45°.

e. What ratio of trigonometry can you use to find the maximum length of the square's side?

The hypotenuse of the triangle created by the diagonal is 6.1804 feet.

You are trying to find either side length s, so use either the sine of 45° or the cosine of 45° to find the side length. Calculations with sine are shown below.

$$\sin 45^\circ = \frac{s}{6.1804}$$

$$6.1804 \bullet \sin 45^\circ = s$$

$$s \approx 4.37 \text{ feet}$$

The maximum side length of the square design should be approximately 4.37 feet.

f. What is the area of the square?

$$A_{\text{square}} = s^2$$

$$A_{\text{square}} = 19.0969 \text{ ft}^2$$

The area of the square is approximately 19.1 ft^2.

g. What is the area of the circle minus the area of the square?

$$30 - 19.1 = 10.9 \text{ ft}^2$$

The tablecloth will have about 10.9 square feet of fabric that is not printed with a square.

G–GMD.1; G–GMD.3★ • Geometry

Volumes of Cylinders, Pyramids, Cones, and Spheres

Common Core State Standards

G–GMD.1

Give an informal argument for the formulas for the circumference of a circle, area of a circle, volume of a cylinder, pyramid, and cone. *Use dissection arguments, Cavalieri's principle, and informal limit arguments.*

G–GMD.3

Use volume formulas for cylinders, pyramids, cones, and spheres to solve problems.★

Problem-Based Task: Cylinders of Sand

The manager of road maintenance for a large city wants to reduce the amount of space needed to store piles of sand. The sand is used on icy roads in the winter. There are 3 piles of sand that are cone-shaped. Each pile has a circumference of 157 feet and a height of 20 feet. The manager is thinking of building a storage area that is in the shape of a cylinder. The cylinder will have the same height as the cone-shaped sand piles, and the cylinder's circumference will be the same as that of the bases of the sand piles. How much area will be saved by building the new cylindrical container?

Problem-Based Task: Cylinders of Sand

Coaching

a. How much area does 1 pile of sand take up?

b. How much area do 3 piles of sand take up?

c. What is the volume of 1 pile of sand?

d. What is the volume of 3 piles of sand?

e. What is the volume of the cylindrical container?

f. How much area does the cylindrical container take up?

g. How much sand is left over after the piles are put into the cylindrical container?

h. How much area will be saved by the new cylindrical container?

Problem-Based Task: Cylinders of Sand

Coaching Sample Responses

a. How much area does 1 pile of sand take up?

The area of the pile of sand is $A = \pi r^2$.

Find the value of r by solving for r in the formula for circumference.

$C = 2\pi r$

$157 = 2\pi r$ Substitute the known value of the circumference.

$\frac{157}{2\pi} = r$ Divide both sides by 2π.

$r \approx 24.987$

The radius of the sand pile is about 24.987 feet.

$A = \pi(24.987^2)$ Substitute the radius value into the area formula.

$A \approx 1961.51$

The area of the base of the pile of sand is 1,961.51 square feet.

b. How much area do 3 piles of sand take up?

$3 \bullet 1961.51 \approx 5884.36$

The area of 3 piles of sand is about 5,884 square feet.

c. What is the volume of 1 pile of sand?

Use the formula for the volume of a cone.

$V = \frac{1}{3} B \bullet h$

$V = \frac{1}{3} \pi r^2 \bullet h$ Substitute the expression for the area of a circle.

$V = \frac{1}{3}(1961.51) \bullet 20$ Substitute the above values for the area of the cone's base and the height of the cone.

$V \approx 13{,}076.67 \text{ ft}^3$

The volume of 1 pile of sand is 13,076.67 cubic feet.

d. What is the volume of 3 piles of sand?

$3 \bullet 13{,}076.67 = 39{,}230.01 \text{ ft}^3$

The volume of 3 piles of sand is about 39,230.01 cubic feet.

e. What is the volume of the cylindrical container?

Use the formula for the volume of a cylinder.

$V = B \bullet h$

$V = 1961.51 \bullet 20 = 39{,}230.20$ cubic feet — Substitute the known area of the base because it has the same dimensions as the base of the cone.

f. How much area does the cylindrical container take up?

It takes up the same area as 1 cone of sand, 1,961.51 square feet.

g. How much sand is left over after the piles are put into the cylindrical container?

There is no sand remaining. All 3 piles fit exactly into the 1 cylinder. The volume of the 3 cones of sand was the same as the volume of the cylinder.

h. How much area will be saved by the new cylindrical container?

The area saved equals the area of 2 piles of sand. The cylinder has the same base area as 1 pile of sand, but now all 3 piles of sand are in one location (the cylinder).

$2 \bullet 1961.51 = 3923.02$

The new cylinder saves approximately 3,923 square feet of space.

This answer makes sense because the formula for the volume of a cone is very similar to the formula for the volume of cylinder, except the cone is one-third the volume of the cylinder. Therefore, 3 piles of sand will fit in 1 cylinder of the same size base and height as the cones of sand.

Common Core State Standard

S–CP.1

Describe events as subsets of a sample space (the set of outcomes) using characteristics (or categories) of the outcomes, or as unions, intersections, or complements of other events ("or," "and," "not").★

Problem-Based Task: Recognizing Equal Sets

William is playing a game that involves rolling dice. On one of his turns he rolls a first die and then a second die. Consider the following events along with the sample space, S.

A: The first die is odd.

B: The sum is odd.

C: There is exactly one odd.

D: The sum is even.

E: The product is odd.

Organize the following ten sets of outcomes into groups of equal sets. Use each set exactly once.

$A, B, C, \overline{D}, E, A \cup E, A \cap E, B \cup D, \overline{B \cap D}, S$

Problem-Based Task: Recognizing Equal Sets

Coaching

a. What is the sample space?

b. What are the outcomes in $\overline{D}$?

c. What are the outcomes in $A \cup E$?

d. What are the outcomes in $A \cap E$?

e. What are the outcomes in $B \cup D$?

f. What are the outcomes in $\overline{B \cap D}$?

g. Now list the ten sets you need to place into groups of equal sets, showing the outcomes in each.

h. Finally, organize the ten sets into groups of equal sets.

Problem-Based Task: Recognizing Equal Sets

Coaching Sample Responses

a. What is the sample space?

(1, 1)	(1, 2)	(1, 3)	(1, 4)	(1, 5)	(1, 6)
(2, 1)	(2, 2)	(2, 3)	(2, 4)	(2, 5)	(2, 6)
(3, 1)	(3, 2)	(3, 3)	(3, 4)	(3, 5)	(3, 6)
(4, 1)	(4, 2)	(4, 3)	(4, 4)	(4, 5)	(4, 6)
(5, 1)	(5, 2)	(5, 3)	(5, 4)	(5, 5)	(5, 6)
(6, 1)	(6, 2)	(6, 3)	(6, 4)	(6, 5)	(6, 6)

b. What are the outcomes in $\overline{D}$?

Note: For convenience, each outcome is written as a pair of single-digit integers. So, 11 represents (1, 1), 12 represents (1, 2), and so on.

$D = \{11, 13, 15, 22, 24, 26, 31, 33, 35, 42, 44, 46, 51, 53, 55, 62, 64, 66\}$

$\overline{D}$ is made up of all the elements in the sample space not found in D.

$\overline{D} = \{12, 14, 16, 21, 23, 25, 32, 34, 36, 41, 43, 45, 52, 54, 56, 61, 63, 65\}$

c. What are the outcomes in $A \cup E$?

$A = \{11, 12, 13, 14, 15, 16, 31, 32, 33, 34, 35, 36, 51, 52, 53, 54, 55, 56\}$

$E = \{11, 13, 15, 31, 33, 35, 51, 53, 55\}$

$A \cup E$ is made up of all the elements that are found in A or E or both.

$A \cup E = A = \{11, 12, 13, 14, 15, 16, 31, 32, 33, 34, 35, 36, 51, 52, 53, 54, 55, 56\}$

d. What are the outcomes in $A \cap E$?

$A = \{11, 12, 13, 14, 15, 16, 31, 32, 33, 34, 35, 36, 51, 52, 53, 54, 55, 56\}$

$E = \{11, 13, 15, 31, 33, 35, 51, 53, 55\}$

$A \cap E$ is made up of all the elements found in A and E.

$A \cap E = E = \{11, 13, 15, 31, 33, 35, 51, 53, 55\}$

e. What are the outcomes in $B \cup D$?

$B = \{12, 14, 16, 21, 23, 25, 32, 34, 36, 41, 43, 45, 52, 54, 56, 61, 63, 65\}$

$D = \{11, 13, 15, 22, 24, 26, 31, 33, 35, 42, 44, 46, 51, 53, 55, 62, 64, 66\}$

$B \cup D$ is made up of all the elements in B or D or both.

$B \cup D = S = \{11, 12, 13, 14, 15, 16, 21, 22, 23, 24, 25, 26, 31, 32, 33, 34, 35, 36, 41, 42, 43, 44, 45, 46, 51, 52, 53, 54, 55, 56, 61, 62, 63, 64, 65, 66\}$

f. What are the outcomes in $\overline{B \cap D}$?

$B = \{12, 14, 16, 21, 23, 25, 32, 34, 36, 41, 43, 45, 52, 54, 56, 61, 63, 65\}$

$D = \{11, 13, 15, 22, 24, 26, 31, 33, 35, 42, 44, 46, 51, 53, 55, 62, 64, 66\}$

$B \cap D = \varnothing$, the empty set.

$\overline{B \cap D}$ is the set of all outcomes that are in the sample space, but not in $B \cap D$.

There are no outcomes in $B \cap D$, so $\overline{B \cap D}$ is the entire sample space.

Therefore, $\overline{B \cap D} = S$.

g. Now list the ten sets you need to place into groups of equal sets, showing the outcomes in each.

$A = \{11, 12, 13, 14, 15, 16, 31, 32, 33, 34, 35, 36, 51, 52, 53, 54, 55, 56\}$

$B = \{12, 14, 16, 21, 23, 25, 32, 34, 36, 41, 43, 45, 52, 54, 56, 61, 63, 65\}$

$C = \{12, 14, 16, 21, 23, 25, 32, 34, 36, 41, 43, 45, 52, 54, 56, 61, 63, 65\}$

$\overline{D} = \{12, 14, 16, 21, 23, 25, 32, 34, 36, 41, 43, 45, 52, 54, 56, 61, 63, 65\}$

$E = \{11, 13, 15, 31, 33, 35, 51, 53, 55\}$

$A \cup E = A$

$A \cap E = E$

$B \cup D = S$

$\overline{B \cap D} = S$

$S = \{11, 12, 13, 14, 15, 16, 21, 22, 23, 24, 25, 26, 31, 32, 33, 34, 35, 36, 41, 42, 43, 44, 45, 46, 51, 52, 53, 54, 55, 56, 61, 62, 63, 64, 65, 66\}$

h. Finally, organize the ten sets into groups of equal sets.

$B = C = \overline{D}$

$A \cup E = A$

$A \cap E = E$

$B \cup D = \overline{B \cap D} = S$

S–CP.2★ • Statistics and Probability

Understanding Independent Events

Common Core State Standard

S–CP.2

Understand that two events *A* and *B* are independent if the probability of *A* and *B* occurring together is the product of their probabilities, and use this characterization to determine if they are independent.★

Problem-Based Task: Tossing Coins

Three events, *A*, *B*, and *C*, are independent if the following conditions are all satisfied:

$$P(A \cap B) = P(A) \bullet P(B)$$

$$P(A \cap C) = P(A) \bullet P(C)$$

$$P(B \cap C) = P(B) \bullet P(C)$$

$$P(A \cap B \cap C) = P(A) \bullet P(B) \bullet P(C)$$

Consider these coin-tossing experiments and associated events:

Experiment 1: Toss a coin 2 times.

- Event *A*: Get heads on the first toss.
- Event *B*: Get heads on the second toss.
- Event *C*: Get exactly one head.

Experiment 2: Toss a coin 3 times.

- Event *A*: Get heads on the first toss.
- Event *B*: Get heads on the second toss.
- Event *C*: Get heads on the third toss.

For each experiment, determine whether events *A*, *B*, and *C* are independent.

Problem-Based Task: Tossing Coins

Coaching

a. For experiment 1, what is the sample space?

b. For experiment 1, what are the probabilities of the individual events *A*, *B*, and *C*?

c. For experiment 1, what are the probabilities of the events $A \cap B$, $A \cap C$, and $B \cap C$?

d. For experiment 1, what is the probability of the event $A \cap B \cap C$? Explain.

e. For experiment 1, are the events *A*, *B*, and *C* independent? Show your work.

f. For experiment 2, what is the sample space?

g. For experiment 2, what are the probabilities of the individual events *A*, *B*, and *C*?

h. For experiment 2, what are the probabilities of the events $A \cap B$, $A \cap C$, and $B \cap C$?

i. For experiment 2, what is the probability of the event $A \cap B \cap C$? Explain.

j. For experiment 2, are the events *A*, *B*, and *C* independent? Show your work.

Problem-Based Task: Tossing Coins

Coaching Sample Responses

a. For experiment 1, what is the sample space?

Sample space = {HH, HT, TH, TT}

b. For experiment 1, what are the probabilities of the individual events *A*, *B*, and *C*?

$P(A)=\frac{2}{4}=\frac{1}{2}$; $P(B)=\frac{2}{4}=\frac{1}{2}$; $P(C)=\frac{2}{4}=\frac{1}{2}$

c. For experiment 1, what are the probabilities of the events $A\cap B$, $A\cap C$, and $B\cap C$?

$P(A\cap B)=\frac{1}{4}$; $P(A\cap C)=\frac{1}{4}$; $P(B\cap C)=\frac{1}{4}$

d. For experiment 1, what is the probability of the event $A\cap B\cap C$? Explain.

$P(A\cap B\cap C)=0$ because there is no outcome with heads first, heads second, and exactly one head. The event $A\cap B\cap C$ is impossible.

e. For experiment 1, are the events *A*, *B*, and *C* independent? Show your work.

No, the events *A*, *B*, and *C* are not independent.

$P(A\cap B\cap C)\neq P(A)\bullet P(B)\bullet P(C)$

$0\neq\frac{1}{2}\bullet\frac{1}{2}\bullet\frac{1}{2}$

f. For experiment 2, what is the sample space?

Sample space = {HHH, HHT, HTH, HTT, THH, THT, TTH, TTT}

g. For experiment 2, what are the probabilities of the individual events *A*, *B*, and *C*?

$P(A)=\frac{4}{8}=\frac{1}{2}$; $P(B)=\frac{4}{8}=\frac{1}{2}$; $P(C)=\frac{4}{8}=\frac{1}{2}$

h. For experiment 2, what are the probabilities of the events $A\cap B$, $A\cap C$, and $B\cap C$?

$P(A\cap B)=\frac{2}{8}=\frac{1}{4}$; $P(A\cap C)=\frac{2}{8}=\frac{1}{4}$; $P(B\cap C)=\frac{2}{8}=\frac{1}{4}$

i. For experiment 2, what is the probability of the event $A \cap B \cap C$? Explain.

$P(A \cap B \cap C) = \frac{1}{8}$ because there are 8 outcomes in the sample space and 1 of them has heads first, heads second, and heads third.

j. For experiment 2, are the events *A*, *B*, and *C* independent? Show your work.

Yes, the events *A*, *B*, and *C* are independent.

$$P(A \cap B \cap C) = P(A) \bullet P(B) \bullet P(C)$$

$$\frac{1}{8} = \frac{1}{2} \bullet \frac{1}{2} \bullet \frac{1}{2}$$

S–CP.3★; S–CP.5★; S–CP.6★ • Statistics and Probability

Introducing Conditional Probability

Common Core State Standards

S–CP.3

Understand the conditional probability of *A* given *B* as *P*(*A* and *B*)/*P*(*B*), and interpret independence of *A* and *B* as saying that the conditional probability of *A* given *B* is the same as the probability of *A*, and the conditional probability of *B* given A is the same as the probability of *B*.★

S–CP.5

Recognize and explain the concepts of conditional probability and independence in everyday language and everyday situations. *For example, compare the chance of having lung cancer if you are a smoker with the chance of being a smoker if you have lung cancer.*★

S–CP.6

Find the conditional probability of *A* given *B* as the fraction of *B*'s outcomes that also belong to *A*, and interpret the answer in terms of the model.★

Problem-Based Task: Does She Have a Brother or a Sister?

Amy has 1 sibling. Find the probability that Amy's sibling is a boy if given the following parameters:

Parameter 1: Nothing is known about the age of Amy or her sibling.

Parameter 2: It is known that Amy is older than her sibling.

Problem-Based Task: Does She Have a Brother or a Sister?

Coaching

a. Restate the problem using conditional probabilities. Assign these variable names to the relevant events: Let B be the event "Amy's sibling is a boy." Let N be the event "Nothing is known about the age of Amy or her sibling." Let AO be the event "Amy is older than her sibling."

b. The problem is about gender. Amy is 1 of 2 children who were born. What is the sample space that shows all possible gender outcomes for the birth of 2 children? Use "b" for boy and "g" for girl.

c. Amy is a girl. What is the reduced sample space that applies to this situation?

d. Assume that the order of letters in each outcome corresponds to birth order. For example, "bg" means that a boy is born and then a girl is born. The first parameter has the condition that nothing is known about the age of Amy and her sibling. Does the sample space {bg, gb, gg} apply to Parameter 1?

e. Find the probability of Parameter 1, $P(B|N)$. Explain your reasoning.

f. Parameter 2 has the condition that Amy is older than her sibling. What is the further-reduced sample space that applies to Parameter 2? Explain your reasoning.

g. Find the probability of Parameter 2, $P(B|AO)$. Explain your reasoning.

Problem-Based Task: Does She Have a Brother or a Sister?

Coaching Sample Responses

a. Restate the problem using conditional probabilities. Assign these variable names to the relevant events: Let *B* be the event "Amy's sibling is a boy." Let *N* be the event "Nothing is known about the age of Amy or her sibling." Let *AO* be the event "Amy is older than her sibling."

Parameter 1 can be restated as "Find $P(B|N)$."

Parameter 2 can be restated as "Find $P(B|AO)$."

b. The problem is about gender. Amy is 1 of 2 children who were born. What is the sample space that shows all possible gender outcomes for the birth of 2 children? Use "b" for boy and "g" for girl.

The sample space is {bb, bg, gb, gg}.

c. Amy is a girl. What is the reduced sample space that applies to this situation?

Since there can't be two boys, the reduced sample space is {bg, gb, gg}.

d. Assume that the order of letters in each outcome corresponds to birth order. For example, "bg" means that a boy is born and then a girl is born. Parameter 1 has the condition that nothing is known about the age of Amy and her sibling. Does the sample space {bg, gb, gg} apply to Parameter 1?

Yes. The sample space consists of the following: a boy being born and then a girl, a girl being born and then a boy, or a girl being born and then another girl.

e. Find the probability of Parameter 1, $P(B|N)$. Explain your reasoning.

$P(B|N)=\frac{2}{3}$ because there are 3 outcomes and 2 of them involve a boy.

f. Parameter 2 has the condition that Amy is older than her sibling. What is the further-reduced sample space that applies to Parameter 2? Explain your reasoning.

{*gb*, *gg*}. If Amy is older, then she must have been born first. Therefore, *bg* is not a possible outcome for Parameter 2.

g. Find the probability of Parameter 2, $P(B|AO)$. Explain your reasoning.

$P(B|AO)=\frac{1}{2}$ because there are 2 outcomes and 1 of them involves a boy.

Common Core State Standards

S–CP.4

Construct and interpret two-way frequency tables of data when two categories are associated with each object being classified. Use the two-way table as a sample space to decide if events are independent and to approximate conditional probabilities. *For example, collect data from a random sample of students in your school on their favorite subject among math, science, and English. Estimate the probability that a randomly selected student from your school will favor science given that the student is in tenth grade. Do the same for other subjects and compare the results.*★

S–CP.5

Recognize and explain the concepts of conditional probability and independence in everyday language and everyday situations. *For example, compare the chance of having lung cancer if you are a smoker with the chance of being a smoker if you have lung cancer.*★

S–CP.6

Find the conditional probability of *A* given *B* as the fraction of *B*'s outcomes that also belong to *A*, and interpret the answer in terms of the model.★

Problem-Based Task: How Many in the Survey?

Some high school students with part-time jobs were asked where they work. The table shows some of the data, with x and y representing missing values.

Grade	Place of employment		
	Taco Heaven	**Burgers Your Way**	**Pizza Palace**
10	1	5	y
11	x	1	5
12	11	3	4

Consider the following events and probabilities.

TEN: The student is in the tenth grade.

ELEVEN: The student is in the eleventh grade.

T: The student works at Taco Heaven.

P: The student works at Pizza Palace.

$$P(T|ELEVEN)=\frac{2}{5}$$

$$P(TEN|P)=\frac{2}{5}$$

Find x, y, and the total number of students surveyed.

Problem-Based Task: How Many in the Survey?

Coaching

a. What is the formula for $P(T|ELEVEN)$ that uses subsets of the sample space?

b. What variable represents the number of outcomes in "*T* and *ELEVEN*"?

c. What expression represents the number of outcomes in "*ELEVEN*"?

d. What is the given value of $P(T|ELEVEN)$?

e. The answer to part a is an equation. Rewrite that equation, substituting the answers to parts b, c, and d.

f. The equation you wrote for part e is a proportion. Solve the proportion.

g. What is the formula for $P(TEN|P)$ that uses subsets of the sample space?

h. What variable represents the number of outcomes in "*TEN* and *P*"?

i. What expression represents the number of outcomes in "*P*"?

j. What is the given value of $P(TEN|P)$?

k. The answer to part g is an equation. Rewrite that equation, substituting the answers to parts h, i, and j.

l. The equation you wrote for part k is a proportion. Solve the proportion.

m. Now that the values of x and y have been calculated, use those values along with the given values in the table to find the total number of students surveyed.

Problem-Based Task: How Many in the Survey?

Coaching Sample Responses

a. What is the formula for $P(T|ELEVEN)$ that uses subsets of the sample space?

$$P(T|ELEVEN) = \frac{\text{number of outcomes in } (T \text{ and } ELEVEN)}{\text{number of outcomes in } (ELEVEN)}$$

b. What variable represents the number of outcomes in "*T* and *ELEVEN*"?

x

c. What expression represents the number of outcomes in "*ELEVEN*"?

$x + 1 + 5$, or $x + 6$

d. What is the given value of $P(T|ELEVEN)$?

$$\frac{2}{5}$$

e. The answer to part a is an equation. Rewrite that equation, substituting the answers to parts b, c, and d.

$$\frac{2}{5} = \frac{x}{x+6}$$

f. The equation you wrote for part e is a proportion. Solve the proportion.

$$\frac{2}{5} = \frac{x}{x+6}$$

$2(x + 6) = 5x$

$2x + 12 = 5x$

$12 = 3x$

$4 = x$

g. What is the formula for $P(TEN|P)$ that uses subsets of the sample space?

$$P(TEN|P) = \frac{\text{number of outcomes in } (TEN \text{ and } P)}{\text{number of outcomes in } (P)}$$

h. What variable represents the number of outcomes in "*TEN* and *P*"?

y

i. What expression represents the number of outcomes in "*P*"?

$y + 5 + 4$, or $y + 9$

j. What is the given value of $P(TEN|P)$?

$\frac{2}{5}$

k. The answer to part g is an equation. Rewrite that equation, substituting the answers to parts h, i, and j.

$\frac{2}{5} = \frac{y}{y+9}$

l. The equation you wrote for part k is a proportion. Solve the proportion.

$\frac{2}{5} = \frac{y}{y+9}$

$2(y + 9) = 5y$

$2y + 18 = 5y$

$18 = 3y$

$6 = y$

m. Now that the values of x and y have been calculated, use those values along with the given values in the table to find the total number of students surveyed.

Sum the values to find the total number of students surveyed.

$1 + 5 + y + x + 1 + 5 + 11 + 3 + 4$

$= 1 + 5 + 6 + 4 + 1 + 5 + 11 + 3 + 4$

$= 40$

40 students were surveyed.

S–CP.7★ • Statistics and Probability

The Addition Rule

Common Core State Standard

S–CP.7

Apply the Addition Rule, $P(A \text{ or } B) = P(A) + P(B) - P(A \text{ and } B)$, and interpret the answer in terms of the model.★

Problem-Based Task: Paying the Bonus?

A university is offering a two-day conference for high school math teachers. Teachers can register for Monday or Tuesday or both days. The conference coordinator compiles the following data:

150 teachers registered for Monday.

95 teachers registered for Tuesday.

210 teachers registered in all.

One registered teacher selected at random will win a laptop computer as a door prize. If the selected teacher has registered for both days, he or she will win an additional bonus prize of $500. What is the probability that the university will have to pay the additional bonus prize?

Problem-Based Task: Paying the Bonus?

Coaching

a. What is the sample space?

b. How many outcomes are in the sample space?

c. Let *M* represent the set of all teachers registered for Monday. What is *P*(*M*), the probability that the selected teacher is registered for Monday?

d. Let *T* represent the set of all teachers registered for Tuesday. What is *P*(*T*), the probability that the selected teacher is registered for Tuesday?

e. You need to find the probability that the university will have to pay the additional bonus prize of $500. The occurrence of what event will require the university to pay the additional bonus?

f. How can you restate the event "registered for both days," using the words "Monday" and "Tuesday"?

g. Using the letters *P*, *M*, and *T*, how can you represent the probability of the event "registered for Monday and Tuesday"?

h. There are 210 teachers registered in all. How can you restate the phrase "registered in all," using the words "Monday" and "Tuesday"?

i. Using the letters *P*, *M*, and *T*, how can you represent the probability of the event "registered for Monday or Tuesday"?

j. What is the probability that the selected teacher is registered for Monday or Tuesday? Explain.

k. What is the rule that relates the probabilities *P*(*M*), *P*(*T*), *P*(*M* or *T*), and *P*(*M* and *T*)?

l. What is the probability that the university will have to pay the additional bonus prize? Show and explain your work.

Problem-Based Task: Paying the Bonus?

Coaching Sample Responses

a. What is the sample space?

The sample space is all the registered teachers.

b. How many outcomes are in the sample space?

210

c. Let *M* represent the set of all teachers registered for Monday. What is *P*(*M*), the probability that the selected teacher is registered for Monday?

$$P(M)=\frac{150}{210}$$

d. Let *T* represent the set of all teachers registered for Tuesday. What is *P*(*T*), the probability that the selected teacher is registered for Tuesday?

$$P(T)=\frac{95}{210}$$

e. You need to find the probability that the university will have to pay the additional bonus prize of $500. The occurrence of what event will require the university to pay the additional bonus?

The university will have to pay the additional bonus if the selected teacher is registered for both days.

f. How can you restate the event "registered for both days," using the words "Monday" and "Tuesday"?

The event can be restated as "registered for Monday and Tuesday."

g. Using the letters *P*, *M*, and *T*, how can you represent the probability of the event "registered for Monday and Tuesday"?

The event can be represented as *P*(*M* and *T*).

h. There are 210 teachers registered in all. How can you restate the phrase "registered in all," using the words "Monday" and "Tuesday"?

The phrase can be restated as "registered for Monday or Tuesday."

i. Using the letters *P*, *M*, and *T*, how can you represent the probability of the event "registered for Monday or Tuesday"?

The event can be represented as $P(M \text{ or } T)$.

j. What is the probability that the selected teacher is registered for Monday or Tuesday? Explain.

$$P(M \text{ or } T)=\frac{210}{210}=1$$

The teacher is selected from the sample space of the 210 registered teachers. All of those registered teachers are registered for Monday or Tuesday or both days, so it is certain that the selected teacher is registered for Monday or Tuesday.

k. What is the rule that relates the probabilities $P(M)$, $P(T)$, $P(M \text{ or } T)$, and $P(M \text{ and } T)$?

The Addition Rule relates the probabilities: $P(M \text{ or } T) = P(M) + P(T) - P(M \text{ and } T)$

l. What is the probability that the university will have to pay the additional bonus prize? Show and explain your work.

$$P(M \text{ or } T) = P(M) + P(T) - P(M \text{ and } T)$$

The addition rule says that you add the probabilities of each event and subtract out the probability of both events occurring. Without subtracting out the intersection of the events, the probabilities of those events are counted twice.

$$\frac{210}{210}=\frac{150}{210}+\frac{95}{210}-P(M \text{ and } T)$$

$$\frac{210}{210}=\frac{245}{210}-P(M \text{ and } T)$$

$$-\frac{35}{210}=-P(M \text{ and } T)$$

$$\frac{35}{210}=P(M \text{ and } T)$$

The probability that the university will have to pay the additional bonus prize is $\frac{35}{210}=\frac{1}{6}\approx 0.17$, or about 17%.

S–CP.8★(+) • Statistics and Probability

The Multiplication Rule

Common Core State Standard

S–CP.8

(+) Apply the general Multiplication Rule in a uniform probability model, $P(A \text{ and } B) = P(A)P(B|A) = P(B)P(A|B)$, and interpret the answer in terms of the model.★

Problem-Based Task: Playing Cards

Kayla is creating a card game. A player in the card game will draw 2 cards from a standard deck of cards, without replacing the first card drawn. Kayla wants sets of cards to be worth more points if they are harder to draw; that is, if the cards have a lower probability of being drawn from the deck. She is comparing drawing cards of a given suit and cards of a given value. Which set of cards is harder to draw: 2 cards that are hearts, or 2 cards that are 8s?

Problem-Based Task: Playing Cards

Coaching

a. What are the suits in a standard deck of cards?

b. How many cards are there of each type of suit?

c. Are the events in the task independent or dependent? Explain your answer.

d. After drawing 1 card that is a heart, what is the probability that the second card is a heart?

e. What is the probability that 2 cards drawn from a deck are hearts?

f. What are the values of the cards in a standard deck?

g. How many of each value are in a deck?

h. After drawing 1 card that is an 8, what is the probability that the second card is an 8?

i. What is the probability that 2 cards drawn from a deck are both 8s?

j. Which set of cards is more difficult to draw: 2 cards that are hearts, or 2 cards that are 8s?

Problem-Based Task: Playing Cards

Coaching Sample Responses

a. What are the suits in a standard card deck?

There are 4 suits in a standard card deck: hearts, diamonds, spades, and clubs.

b. How many cards are there of each type of suit?

A deck of cards has 52 cards. There are an equal number of cards in each of the 4 suits, so there are $52 \div 4 = 13$ cards of each suit.

c. Are the events in the task independent or dependent? Explain your answer.

The events are dependent. The probability of choosing a card that is a heart the second time is affected by what card is drawn the first time. If a heart is drawn the first time, then there is now 1 less heart card in the total number of outcomes. The same is true for drawing an 8 on the second draw if an 8 is drawn the first time.

d. After drawing 1 card that is a heart, what is the probability that the second card is a heart?

After 1 card is drawn, there are $52 - 1 = 51$ cards remaining in the deck. There are 13 cards of each suit. After 1 heart is drawn, there are: $13 - 1 = 12$ hearts remaining. To find the probability that the second card is a heart given that the first card is a heart, divide the number of hearts remaining by the total cards remaining.

Let *FH* represent drawing a heart for the first card.

Let *SH* represent drawing a heart for the second card.

$$P(SH|FH) = \frac{\text{cards remaining that are hearts}}{\text{cards remaining in deck}}$$

$$P(SH|FH) = \frac{12}{51} = \frac{4}{17}$$

e. What is the probability that 2 cards drawn from a deck are both hearts?

There are two events: *FH*, the first card drawn is a heart, and *SH*, the second card drawn is a heart. The second event is dependent on the first. The Multiplication Rule can be used to find the probability that both events occur.

$$P(A \text{ and } B) = P(A) \bullet P(B|A)$$

Substitute *FH* for *A* and *SH* for *B*.

$$P(FH \text{ and } SH) = P(FH) \bullet P(SH|FH)$$

Find *P*(*FH*), the probability of the first drawn card being a heart.

$$P(FH) = \frac{\text{number of hearts}}{\text{total number of cards}}$$

$$P(FH) = \frac{13}{52} = \frac{1}{4}$$

From part d we determined $P(SH|FH)$, the probability of the second drawn card being a heart given that the first drawn card is a heart.

$$P(SH|FH) = \frac{4}{17}$$

Substitute the found probabilities into the equation for the Multiplication Rule and solve.

$$P(FH \text{ and } SH) = P(FH) \bullet P(SH|FH) = \frac{1}{4} \bullet \frac{4}{17} = \frac{1}{17} \approx 0.059$$

f. What are the values of the cards in a standard deck?

The values are: 2, 3, 4, 5, 6, 7, 8, 9, 10, jack, queen, king, and ace.

g. How many of each value are in a deck?

There are 13 values, and an equal number of each value in the deck.

There are 52 ÷ 13 = 4 cards of each value.

h. After drawing 1 card that is an 8, what is the probability that the second card is an 8?

After 1 card is drawn, there are 52 – 1 = 51 cards remaining in the deck. There are 4 cards of each value. After 1 card that is an 8 is drawn, there are 3 cards that are 8s remaining. To find the probability that the second card is an 8 given that the first card is an 8, divide the number of 8s remaining by the total cards remaining.

Let *FE* represent drawing an 8 for the first card.

Let *SE* represent drawing an 8 for the second card.

$$P(SE|FE) = \frac{\text{cards remaining that are 8s}}{\text{cards remaining in deck}}$$

$$P(SE|FE) = \frac{3}{51} = \frac{1}{17} \approx 0.059$$

i. What is the probability that 2 cards drawn from a deck are both 8s?

There are two events: *FE*, the first card drawn is an 8, and *SE*, the second card drawn is an 8. The second event is dependent on the first. The Multiplication Rule can be used to find the probability that both events occur.

$$P(A \text{ and } B) = P(A) \bullet P(B|A)$$

Substitute *FE* for *A* and *SE* for *B*.

$$P(FE \text{ and } SE) = P(FE) \bullet P(SE|FE)$$

Find *P*(*FE*), the probability of the first drawn card being an 8.

$$P(FE) = \frac{\text{number of 8s}}{\text{total number of cards}}$$

$$P(FE) = \frac{4}{52} = \frac{1}{13} \approx 0.077$$

From part h we determined $P(SE|FE)$, the probability of the second drawn card being an 8 given that the first drawn card is an 8.

$$P(SE|FE) = \frac{1}{17}$$

Substitute the found probabilities into the equation for the Multiplication Rule and solve.

$$P(FE \text{ and } SE) = P(FE) \bullet P(SE|FE) = \frac{1}{13} \bullet \frac{1}{17} = \frac{1}{221} \approx 0.0045$$

j. Which set of cards is more difficult to draw: 2 cards that are hearts, or 2 cards that are 8s?

The set that has the lower probability is less likely, and is more difficult to draw.

Compare the probability of *P*(*FH* and *SH*) with the probability of *P*(*FE* and *SE*).

$$0.059 > 0.0045$$

The probability of drawing 2 cards that are 8s is less than the probability of drawing 2 cards that are hearts, so drawing 2 cards that are 8s is more difficult.

S–CP.9★(+) • Statistics and Probability

Combinations and Permutations

Common Core State Standard

S–CP.9

(+) Use permutations and combinations to compute probabilities of compound events and solve problems.★

Problem-Based Task: On the Court

There are 5 different positions on a basketball team: point guard, shooting guard, power forward, center, and small forward. In professional basketball, only 12 different players per team are allowed to be ready to play in a game, with 5 allowed on the court at any time. If any of the 12 team members can play any of the 5 positions, how many different ways can the team be positioned on the court?

Problem-Based Task: On the Court

Coaching

a. Are combinations or permutations being counted?

b. Can there be repetition within the groups?

c. How many positions are there?

d. How many choices are there for each position?

e. How many different ways can the team be positioned?

Problem-Based Task: On the Court

Coaching Sample Responses

a. Are combinations or permutations being counted?

There are 5 unique positions, so the arrangement of the people in the group matters. Since order matters, permutations are being counted.

b. Can there be repetition within the groups?

Each position is taken by 1 player. Since 1 player cannot be in more than 1 position, there can't be repetition.

c. How many positions are there?

There are 5 different positions.

d. How many choices are there for each position?

Position	Number of players who can be selected	Explanation
1	12	Any of the 12 players can be selected.
2	11	1 player has been selected, so 12 – 1 = 11 players remain.
3	10	2 players have been selected, so 12 – 2 = 10 players remain.
4	9	3 players have been selected, so 12 – 3 = 9 players remain.
5	8	4 players have been selected, so 12 – 4 = 8 players remain.

e. How many different ways can the team be positioned?

Find the product of the number of players for each position.

$12 \bullet 11 \bullet 10 \bullet 9 \bullet 8 = 95{,}040$

A graphing calculator can also be used.

On a TI-83/84:

Step 1: In a blank window, type [12], the value of *n*.

Step 2: Press the [MATH] key.

Step 3: Arrow over to PRB to access the probability menu.

Step 4: Arrow down to 2: nPr. Press [ENTER].

Step 5: Input the value of *r*, which is [5], and press [ENTER]. Note this counts the number of permutations without repetition.

On a TI-Nspire:

Step 1: From the calculate window, press the [menu] key.

Step 2: Arrow down to 5: Probability.

Step 3: Arrow down to 2: Permutations.

Step 4: Where the screen says "nPr(", type [12][,][5], and press enter. Note this counts the number of permutations without repetition.

There are 95,040 ways the team can be positioned.

S–CP.9★(+) • Statistics and Probability

Probability with Combinatorics

Common Core State Standard

S–CP.9

(+) Use permutations and combinations to compute probabilities of compound events and solve problems.★

Problem-Based Task: The Apple Game

Luis and Rose are playing a word game on the whiteboard. Luis has drawn an apple tree with apples on it. He has chosen a secret word and written blanks for each letter of the word. Rose must guess the letters of the word one at a time; if she's correct, Luis will fill in the letter she guessed. If she's incorrect, Luis will erase one of the apples on the tree. Rose keeps guessing letters until she knows what the word is, or until all the apples have been erased.

Luis's secret word is *young*. Rose decides to randomly guess a consonant first, then a vowel. What is the probability that both selected letters are in Luis's word? Let Y be a consonant.

Problem-Based Task: The Apple Game

Coaching

a. What is the probability being determined?

b. What is the size of the sample space?

c. How many different ways can Rose correctly guess the 2 letters?

d. What is the probability that Rose correctly selects the consonant and the vowel?

Problem-Based Task: The Apple Game

Coaching Sample Responses

a. What is the probability being determined?

The probability being determined is whether Rose's randomly guessed consonant and vowel are in Luis's word. Let *CV* represent the event of correctly guessing a consonant and vowel: $P(CV)$.

b. What is the size of the sample space?

The sample space is all of the ways Rose can guess a consonant and a vowel.

If Y is considered a consonant, there are 21 consonants and 5 vowels in the alphabet.

Rose can guess any of the consonants in her first guess and any of the vowels in her second guess.

The number of combinations is the product of the number of consonants and the number of vowels.

$21 \bullet 5 = 105$

c. How many different ways can Rose correctly guess the 2 letters?

Luis's secret word is *young*, which has 3 consonants (Y, N, and G) and 2 vowels (O and U).

Any of the 3 consonants can be the first guess, and any of the 2 vowels can be the second guess.

The number of ways Rose can correctly guess 2 letters in the word is the product of the number of consonants in the word and the number of vowels in the word.

$3 \bullet 2 = 6$

d. What is the probability that Rose correctly selects the consonant and the vowel?

The probability of an event is $\frac{\text{number of ways event can occur}}{\text{total number of events}}$.

Use the determined information to find $P(CV)$, the probability of correctly guessing a consonant and a vowel.

Divide the number of ways correct letters can be guessed by the total number of ways the letters can be guessed.

$$P(CV) = \frac{\text{number of ways Rose can guess correctly}}{\text{number of ways a consonant and a vowel can be guessed}}$$

$$P(CV) = \frac{6}{105} = \frac{2}{35} \approx 0.057$$

The probability that both selected letters are in Luis's word is 0.057.

S–MD.6★(+) • Statistics and Probability

Making Decisions

Common Core State Standard

S–MD.6

(+) Use probabilities to make fair decisions (e.g., drawing by lots, using a random number generator).★

Problem-Based Task: Two of a Kind

Shelby is creating a new card game. There are 10 cards in a stack. There are 5 different numbers, 1 through 5, and there are 2 of each number in the stack. The cards are shuffled. A player selects 2 cards from the stack. If the 2 cards have the same value, the player wins. If not, the cards are returned to the stack and reshuffled. Then, the next player selects 2 cards. The first player to select a pair wins. Is the game fair? Do any additional rules need to be added to improve the game's fairness?

Problem-Based Task: Two of a Kind

Coaching

a. How do you win the game?

b. What is the probability that the first player wins?

c. If the first player doesn't win, what is the probability that the second player wins?

d. Does the order of the players impact a player's likelihood of winning?

e. How can you improve the game's fairness?

Problem-Based Task: Two of a Kind

Coaching Sample Responses

a. How do you win the game?

You win the game by selecting 2 cards with the same value. For example, if you select a 2 and then another 2, you win the game.

b. What is the probability that the first player wins?

To win, the second card must match the first. The first card can be any value. After selecting the first card, there are 10 – 1 = 9 cards remaining in the stack. Only 1 of these cards will have the same value as the first. The probability that this card is selected is the probability that the first player wins:

$$P(\text{first player wins}) = P(\text{second card has the same value as first})$$

$$P(\text{first player wins}) = \frac{\text{cards in deck with same value as first}}{\text{cards remaining in deck}}$$

$$P(\text{first player wins}) = \frac{1}{9}$$

c. If the first player doesn't win, what is the probability that the second player wins?

If the first player doesn't win, the selected cards are replaced in the stack and reshuffled. The probability that the second player wins is the probability that the second card selected has the same value as the first.

$$P(\text{second player wins}) = P(\text{second card has the same value as first})$$

$$P(\text{second player wins}) = \frac{\text{cards in deck with same value as first}}{\text{cards remaining in deck}}$$

$$P(\text{second player wins}) = \frac{1}{9}$$

d. Does the order of the players impact a player's likelihood of winning?

If the first player wins, then no other players get to play. Similarly, if the second player wins, the third player and any remaining players also don't get a chance to play. Each player has the same likelihood of winning, since the probability that a player will select a pair is the same. However, any players whose turn follows that of a winning player automatically lose.

e. How can you improve the game's fairness?

The selection of the cards is a fair way to select a winner, since selecting a pair (and therefore, winning) has the same probability for each player. To improve the game's fairness, the order of the players should also be fair. The order should be randomly selected using a random number generator. Ways to randomly select the order include rolling a die, selecting numbers or names from a hat without looking, or selecting shuffled cards with different values.

The game is fair, based on the probability of selecting a winning hand, but could be improved by randomly selecting the order of the players.

NAME:

S–MD.7★(+) • Statistics and Probability

Analyzing Decisions

Common Core State Standard

S–MD.7

(+) Analyze decisions and strategies using probability concepts (e.g., product testing, medical testing, pulling a hockey goalie at the end of a game).★

Problem-Based Task: Changing Channels

A radio station conducted research to learn more about its listeners. The research shows that the song played just before a commercial break has an impact on whether or not listeners change the radio station. The station manager wants to keep as many listeners tuned during the commercials, to encourage advertisers to buy commercials.

The station manager is trying to determine which type of song the deejay should play just before a commercial break. The station plays pop, dance, and contemporary music. The probability that a listener will change the channel during any type of song is in the table below.

Song type	Probability of changing the channel	Probability of still listening
Pop	0.21	0.79
Dance	0.18	0.82
Contemporary	0.28	0.72

The probability that a listener will change the channel during the beginning of the commercial break depends on the song just prior to the commercial, as shown below.

Song type prior to commercial	Probability of changing the channel during the commercial break	Probability of still listening
Pop	0.55	0.45
Dance	0.60	0.40
Contemporary	0.32	0.68

Which type of song should the deejay play just prior to the commercial, to keep the most listeners during the commercial break? Assume that the probabilities take into consideration any new listeners.

Problem-Based Task: Changing Channels

Coaching

a. When could a listener change the channel?

b. Create a table showing all the paths a listener could take during the song just prior to the commercial and during the commercial.

c. What is the probability that a listener will still be tuned to the radio station after each chosen path?

d. Which type of song should the deejay play prior to a commercial in order to keep as many listeners as possible during the commercial break?

Problem-Based Task: Changing Channels

Coaching Sample Responses

a. When could a listener change the channel?

A listener could change the channel during a song or during a commercial break.

b. Create a table showing all the paths a listener could take during the song just prior to the commercial and during the commercial.

There are three different song types. The listener could stop listening during the song. If the listener is still tuned in after the song, the listener could stop listening during the commercial.

Song	Still listening during song?	Still listening during commercial?
Pop	Y	Y
Pop	Y	N
Pop	N	N
Dance	Y	Y
Dance	Y	N
Dance	N	N
Contemporary	Y	Y
Contemporary	Y	N
Contemporary	N	N

c. What is the probability that a listener will still be tuned to the radio station after each chosen path?

For each path, write the probability at each point that a listener will either still be tuned in or not. Recall that the probability that a listener will change the channel during the commercial depends on the prior song choice. If a listener is not listening after the song, there is a probability of 0 that the listener will still be listening during the commercial.

Song	Still listening during song?	Probability	Still listening during commercial?	Probability
Pop	Y	0.79	Y	0.45
Pop	Y	0.79	N	0.55
Pop	N	0.21	N	0
Dance	Y	0.82	Y	0.40
Dance	Y	0.82	N	0.60
Dance	N	0.18	N	0
Contemporary	Y	0.72	Y	0.68
Contemporary	Y	0.72	N	0.32
Contemporary	N	0.28	N	0

There is only one path for each song type where the listener is still listening during the commercial. Find the product of the probability that the listener is still listening after the song and during the commercial.

P(still listening during the pop song and during the commercial) = 0.79 • 0.45 = 0.356

P(still listening during the dance song and during the commercial) = 0.82 • 0.40 = 0.328

P(still listening during the contemporary song and during the commercial) = 0.72 • 0.68 = 0.490

d. Which type of song should the deejay play prior to a commercial in order to keep as many listeners as possible during the commercial break?

The probability that a listener is still listening is greatest if the song played prior to the commercial is a contemporary song. Therefore, the deejay should play a contemporary song just prior to a commercial.